MW01108231

Three Prehistoric Inventions That Shaped Us

PETER LANG
New York • Washington, D.C./Baltimore • Bern
Frankfurt • Berlin • Brussels • Vienna • Oxford

DAVID MARTEL JOHNSON

Three Prehistoric Inventions That Shaped Us

PETER LANG
New York • Washington, D.C./Baltimore • Bern
Frankfurt • Berlin • Brussels • Vienna • Oxford

Library of Congress Cataloging-in-Publication Data

Johnson, David Martel.
Three prehistoric inventions that shaped us / David Martel Johnson.
p. cm.
Includes bibliographical references and index.
1. Human evolution. 2. Social evolution. 3. Behavior evolution.
4. Language and culture. 5. Religion and culture.
6. Animals, Domestic. I. Title.
GN281.J59 599.93'8—dc22 2010029727
ISBN 978-1-4331-1090-0

Bibliographic information published by **Die Deutsche Nationalbibliothek**.
Die Deutsche Nationalbibliothek lists this publication in the "Deutsche Nationalbibliografie"; detailed bibliographic data is available on the Internet at http://dnb.d-nb.de/.

Painting on cover by Kirsten Johnson

The paper in this book meets the guidelines for permanence and durability of the Committee on Production Guidelines for Book Longevity of the Council of Library Resources.

Printed in Germany

This book is dedicated to Gretchen Emily Johnson, my daughter who has Down Syndrome and very little language. She has helped me see that being human is more than, and in some ways surprisingly different from, what many wise people suppose.

Contents

Preface: Towards a More Complete Scientific Picture of our Species and its History

> How do we in fact differ today from our ancient ancestors of sixty thousand years ago? The essential transformations that have taken place since then are not . . . to be situated in the genetically inherited structure of the organism. They reside, rather, in the world into which a child is born and within which the child comes to take its place, and in the learned relationships between that newborn child and that emerging world. That may be described as a cultural world, and it is one that generations of earlier people have helped shape.
>
> Colin Renfrew, *Prehistory*, p.90

> If you are not able to laugh at yourself, somebody else will do the job for you.
>
> Proverbial saying

This book is both a sequel and a prequel to another book I published in 2003 with Open Court Press, Chicago called, *How History Made the Mind: the Cultural Origins of Objective Thinking*. One reason for calling the volume now before you a sequel, is that it corrects at least one important mistake that I made in that earlier text. The mistake is this: At the time of writing the previous text, I had formed the impression that, because of the influence of mass media, it no longer was possible to find any people on earth who did not possess modern human minds that were products of the two, large-scale cultural events described and discussed in that book. The first of those events was the "Upper Paleolithic Revolution" that seemed to have been closely associated with humans' development of sophisticated, syntactically organized language—something that, in the opinion of many present-day archeologists, took place between 60,000 and 30,000 years ago. The second, more recent of the two events was what I called the "Greek revolution"—i.e. the invention during early historical times, about 3,000 years before the present, of consistently non-mythological thinking and of the set of mental techniques we now usually refer to as "reason." However, after publication of the first book, further reading and reflection convinced me that it was a mistake to suppose that all people now alive were influenced by both of those events. Accordingly, in one section (4.3) of this text, I have quite a lot to say about certain people—members of the Pirahã and Mundurukú aboriginal tribes of Brazil—of a kind whose existence I once denied.[1]

That is, it now seems to me that the members of those tribes think in ways that were influenced by the Upper Paleolithic Revolution, since we know from observation that they speak genuine, syntactically organized human languages. On the other hand, however, carefully controlled tests administered by visiting anthropologists and psychologists have established that they also are not able to count in a competent and reliable fashion beyond the number three (in the first case) and beyond the number five (in the second case). This lack of counting skills apparently indicates that the Pirahãs and Mundurukús are not capable of thinking about things in the same precise, rational, literal, and non-mythological ways that are characteristic of the great majority of the human beings who now live on our planet. Thus it seems appropriate to conclude that those same people somehow missed being influenced by the invention of reason that began and then spread outwards from one or several locations on the Greek mainland beginning about 3000 years ago.

The book also counts as a prequel ("a preparation after the fact") to the earlier one, because its project is to search further back in time, in order to cast light on what the sources were, from which the Upper Paleolithic Revolution—and together with it, sophisticated language—finally were able to arise. In particular, I shall argue in the following pages that language of this sort (and the kind of thinking that is associated with it) did not "appear out of the blue," in the sense of having been nothing more than inherited products of the genes shared by all of the members of our species, because archeological investigations have shown in a fairly convincing way that language of this sort came into existence at several different places that were separated from one another by thousands of miles, and also occurred at different times that were separated from one another by thousands of years. Events determined just by biology—for example, a child's development in the womb of two legs and two arms, its becoming either a male or female, its eventually growing hair, its "learning" to walk, and its entering into the stage of life known as puberty—do not act in anything like this same way. For example, biological events characteristically happen to every single undamaged member of a species, and not just to members of particular and separate, culturally determined groups.

---

[1] Still another factor that forced me to begin thinking about humans in more inclusive terms than I had done before, was experience with my daughter Gretchen, who has Down syndrome, and to whom I have dedicated this book.

In my view, humans' development of sophisticated language was largely cultural in character, and therefore it was necessary for it to be prepared for by at least two earlier cultural developments. As I shall explain further in the following chapters, I believe the revolutionary development that most closely preceded the acquisition of language, was our species' first domestication of animals. In my opinion, the kind of non-human animals our ancestors learned to domesticate before any other such creatures were wolves or dogs. Furthermore, those animals probably succeeded in teaching their human "masters" certain lessons about how they ought to organize (i) their own lives, (ii) their own societies, and (iii) their general habits of thinking. To be still more specific, wolves provided our ancestors with a practical example that showed them how useful it was to live, and also to contemplate experienced objects, events, and facts, in terms of a strict hierarchy. Later a hierarchy of much the same kind proved to be either similar to, or identical with, the special sort of thinking that is presupposed by the forms, structures, and/or rules of linguistic syntax.

I intend to argue that another, still earlier, and even more basic cultural development that prepared humans for the acquisition of language was what we might call our ancestors' invention of religious consciousness. My reason for believing that this also was a necessary step towards acquiring the kind of human nature we now possess, is that it allowed and encouraged humans—for the first time—to draw a fairly sharp mental line between themselves on one side, and everything else they saw, heard, and felt on the other. Accordingly, sophisticated language (as opposed to ways of communicating that were nothing more than extensions of "body language,") were not able to get started, in the absence of some such distinction, because language of the type we now are discussing is mostly abstract and general, as opposed to concrete and particular. Thus it also was different in this respect from everything else that our ancestors encountered in the world around them.[2]

---

[2] When I mentioned to my dentist that I was writing a book entitled *Three Prehistoric Inventions that Shaped Us*, she said: "What do you mean—something like the wheel?" It is an undeniable fact that the wheel was an important prehistoric invention that some past humans made. But I have not chosen either it or any other invention of the same sort to be the focus of this book because of the fact that, although it influenced and changed many of our ways of doing things, as well as many of the items we then brought into existence (e.g. it led us to substitute roads for paths, and wagons for

At the present time, one popular approach to roughly these same matters—an approach to which I am opposed—is the one illustrated by what many people call the "cognitive revolution."[3] This phrase refers to a rationalistic epistemology or theory of knowledge that appeared within the living memory of people who are as old as I am. As many commentators have pointed out, this special type of theory has some claim to be more firmly based on detailed results of scientific investigations than all previous forms of rationalism, like those adopted by Plato, Descartes, Spinoza, and Leibniz. The basic idea of the cognitive revolution is that human beings like us already have a great deal of inborn, "instinctual" knowledge, which acts as a background, presupposition, and facilitator for everything we subsequently succeed in learning by the use of our reasoning and senses. In opposition to what the behaviorist psychologist B.F. Skinner called "the group mind," Noam Chomsky and other representatives of the cognitive revolution say that the members of each individual species only learn those pieces of knowledge that are particularly appropriate to them, and that they also learn that epistemological material in a distinctive fashion, which also happens to be characteristic of that species. For instance, Chomsky has maintained over the course of most of his career, that no other creatures besides the humans of our species are capable of developing fully-fledged language, because their brains are the only ones that are organized in such a way as to make acquisition of language of that sort possible.

I am willing to grant the general point that the ideas that launched the cognitive revolution are positive and fairly commendable ones that have brought us closer to obtaining an accurate theoretical conception of what it means to be human. Nevertheless, I also believe that those ideas leave something out of account that scientifically inspired theorists need to recognize, if they really do intend to clarify and make sense of what present-day human nature is and how it works. To be more specific, most of the present-day forms of those epistemological positions that we refer to as empiricism and rationalism (including the Chomskean theory) are "mechanistic-like" accounts of human thinking that do not make any serious reference to factors included in the cultural history of human beings—e.g., hu-

---

sledges), it did not also did not have the effect of shaping us ourselves—i.e. our shared human nature—in any direct fashion.

[3] Several main aspects of this revolution are illustrated and discussed in Johnson and Erneling, 1997.

mans' personal creativity, including the special sort of freedom they possess and their power to choose. For example, Chomsky and like-minded individuals show by what they say that they are strongly prejudiced in favor of the methods and findings of the natural sciences, and against those of the social sciences (with the sole possible exception of the science of linguistics). Because of that fact, these people also reject the validity of various discoveries that have been made by representatives of humanistic disciplines like history and archeology, because they consider all such so-called discoveries to be nothing more than instances of what Daniel Dennett calls (see 2006, p.103) "premature curiosity satisfaction." In my view, however, this dismissive attitude has turned out to be a disastrous mistake, because of the fact that it ignores the explanatory power associated with observationally justified statements like the one I quoted from the prehistorian Colin Renfrew at the beginning of this preface.

An alternative means of describing the problem towards which I now am pointing is to say that, while Chomsky and like-minded individuals have abandoned the old-fashioned, Aristotelian doctrine that all knowledge, except for logic and mathematics, must be obtained from direct sensory experience, they continue to affirm the truth of another, equally troublesome dogma that Aristotle once affirmed. That second dogma is that every legitimate piece of scientific knowledge has to be something that is precise, necessary, and timelessly true. However, the problem with this last idea is that many historical and cultural facts about human intellectual development that are important and scientifically useful to us, are both vague and merely contingent, and that they only make sense in the context of certain specific times and places, rather than in all times and places. For this reason, I suggest that a systematic policy of ignoring all discoveries made on the basis of historical and archeological research—which the cognitive revolution evidently recommends—is bound to prevent us from obtaining scientific understanding of many basic facts about past and present human beings.

Quite a few other authors (see e.g. Philip Lieberman, 1991 and 2000) already have provided their readers with a great deal of information about the physiological bases of human nature. And therefore I feel justified in having largely devoted both of my two authored books to the comparatively more neglected topic of the historical and cultural sources that have created, and which still continue to stand behind that nature. Accordingly, my project here is to adjudicate between two apparently irreconcilably viewpoints—namely, the view taken by historians who do not know very much, or care very much, about human biology and physiology, and the viewpoint of philosophers and scientists who neither know nor care very much about humans' cultural history. The fundamental point for which I shall argue in this book is that the best, most adequate, and most explanatory account of human nature, which all modern investigators presumably are seeking to find, is divided from itself in those two warring camps.

To summarize, then, what were the past cultural revolutions that made us into the kind of creatures we are now? My two authored books, working together, defend the answer that there were at least four such events. They were—in temporal order—(i) the invention of religious consciousness, (ii) the domestication of animals (more specifically, dogs), (iii) the invention of syntactical language, and (iv) the invention of Greek-like reason. As far as I can tell at the present time, all the undamaged members of our species who are now alive have been influenced by the first three of these occurrences. But—as observation now has shown—there remain at least some people in the world today (e.g. the Pirahãs) who still have not been touched by the fourth.

I want to thank all those people who have helped me to write, and finally to publish, this book, over the course of the last seven years. For instance, Oksana Silkina did a great deal of the computer work for the preparation of the book that I was not competent to do myself. Also, I owe much to many of the students in my classes, who witnessed—and did their best to learn from—my first clumsy attempts to get my ideas clear. Still more particularly, my students Jason Kennedy and Martin Veser, and my graduate assistant Ross Sweeny, provided me with helpful "feed-back" from their detailed consideration of some of the first drafts of chapters. I also want to thank my professional artist daughter, Kirsten, who produced the wonderful cover art for the book, and my equally wonderful granddaughter Sara Sofia Johnson, who served as Kirsten's model. Again, and as usual, I owe a special debt to my wife Barbara, who somehow managed to keep me, our family, and herself both sane and going strong, despite the problem of having a writer for a husband.

Acknowledgements

Page ix: Excerpt from p. 90 of Colin Renfrew, *Prehistory: the Making of the Modern Mind*, New York, NY, The Modern Library, © 2007 by Colin Renfrew, 2008. Reprinted by permission of Weidenfeld & Nicholson, an imprint of The Orion Publishing Group, London. All rights reserved.

Pages 33-4: Excerpt from pp. 21-2 of Terrence W. Deacon, *The Symbolic Species: The Co-evolution of Language and the Brain*, New York, NY, W.W. Norton & Company, © 1997 by Terrence W. Deacon, 1997. Reprinted by permission of the publisher. All rights reserved.

Page 56: Excerpt from pp.138-9 of Ian Tattersall, *Becoming Human: Evolution and Human Uniqueness*, New York, NY, Harcourt Brace & Company, © 1998 by Ian Tattersall, 1998. Reprinted by permission of Houghton Mifflin Harcourt Publishing Company. All rights reserved.

Pages 58-9: Excerpts from Lynda Hurst, "Sprinting down the evolutionary highway," a short report that appeared in *Toronto Star* newspaper, Sunday, February 3, 2008, p. ID4, Toronto, Canada, © 2008 Torstar Syndication Services, 2008. Reprinted by permission of the publisher. All rights reserved.

Page 64: Excerpt from p. 15 of "A Love Story: Our Bond with Dogs" by Angus Phillips, *National Geographic*, January 2002, pp.12-31. Reprinted by permission of the publisher. All rights reserved.

Pages 68-71: Short article, "A Murder of Crows, written by Rebecca Dube, from *The Globe and Mail* newspaper, Friday Dec. 14, 2007, pp. L1-2, Toronto, Canada, © iCopyright Inc., 2007. Reprinted by permission of the publisher. All rights reserved.

Page 77: Excerpt from p. 5 of Derek Bickerton, *Language & Species*, Chicago IL, University of Chicago Press, © 1990 by The University of Chicago, 1992. Reprinted by permission of the publisher. All rights reserved.

Page 87: Figure and caption from p.497 of Peter Gordon, "Numerical Cognition without Words: Evidence from Amazonia," *Science*, Vol.306 (October 15, 2004), © 2004 American Association for the Advancement of Science, publisher, 2004. Reprinted by permission of the publisher. All rights reserved.

Page 90 note 46: Excerpt from p.4 of Candace Savage, *Bird Brains*, Vancouver, Canada, Greystone Books: an imprint of D&M Publishers Inc., © 1995 by Candace Savage, 1995. Reprinted by permission of the publisher. All rights reserved.

Page 110: Excerpt from p.12 of Helen Keller, *The Story of my Life*, Mineola, NY, Dover Publications Inc., © 1996 by Dover Publications Inc., 1996. Reprinted by permission of the publisher. All rights reserved.

Page 113: Excerpt from pp.88-93 of Candice Savage, *Crows: Encounters with the Wise Guys of the Avian World*, Vancouver, Canada, Greystone Books: an imprint of D&M Publishers Inc., © 2005 by Candace Savage, 2005. Reprinted by permission from the publisher. All rights reserved.

Pages 115-6: Excerpt from p.169 of Noam Chomsky, *Language and the Problems of Knowledge: The Managua Lectures*, Cambridge, MA, MIT Press, © 1988 Massachusetts Institute of Technology, 2001. Reprinted by permission of the publisher. All rights reserved.

Pages 125-6: Excerpt from p.1 of Walter Burkert, *Creation of the Sacred: Tracks of Biology in Early Religions*, Cambridge, MA, Harvard University Press, © by the President

Chapter 1

The Oldest Question: What Separates Human Beings from All Other Creatures?

> Facts are ventriloquists' dummies. Sitting on a wise man's knee they may be made to utter words of wisdom; elsewhere, they say nothing, or talk nonsense, or indulge in sheer diabolism.
>
> Aldous Huxley

1.1 Hummingbirds and Homo Sapiens

> Human action can be modified to some extent, but human nature cannot be changed.
>
> Abraham Lincoln

> The only world we can understand is the one we have created for ourselves.
>
> Friedrich Nietzsche

One of the jobs people need to do when they arrive for a visit to our family cottage, located on an island in a lake north of Toronto, is to clear away spiders' webs from the hummingbird feeder hanging on the front porch, take the feeder inside, wash it, refill it with hummingbird food—water with dissolved sugar—then hang it outside again. Whenever I go through that procedure (or as my wife would say, whenever I watch her doing it, from a comfortable seat in front of the computer), the same thought always occurs to me. I find myself thinking: "What a silly, unrealistic little thing that feeder is, as proved by the fact that the people who designed it apparently didn't know what hummingbirds were like." Let me elaborate. The feeder to which I am referring is a round flat plastic object, about the size of a small bread plate, with a thin metal bar ending in a hook, rising from its center. It also has a scalloped edge that, when viewed from above, makes it look like a flower. The detachable cover on top is bright red in color, while the bottom part that holds the feed is made of clear plastic. What bothers me about this object is neither the red top (hummingbirds need bright colors in order to find the feeder and be attracted to it), nor its whimsical, flower-shaped edge. The problem is that the feeder has six evenly spaced feeding holes, each one associated with its own individual perch located along the scalloped edge. This feature strikes me as ridiculous, because it is at odds with something that anyone who watches hummingbirds feeding for more than fifteen minutes ought to know—namely, that these birds always insist on eating alone, and threaten, bully, and chase each other, until only one of them remains to have access to the food.

Here are a couple of confessions: First, I admit that the point just mentioned is not entirely true since, on a few occasions, each lasting for at least a minute, I have been lucky enough to see two hummingbirds calmly sipping sugar water from the same feeder at the same time. My speculation about those cases (whose truth I cannot prove one way or the other) is that they involve mated pairs. Second, I also admit that I do not literally believe the individual or individuals who designed the feeder did not know anything about hummingbirds. Rather, a more likely hypothesis is that he, she, or they designed the object the way they did, in order to sell it to naïve people like me of twenty-five years ago. In other words, they made it attractive to individuals who, not yet having had a chance to observe hummingbirds up close, were not yet in a position to decide what feeder to buy, on the basis of concrete knowledge about the birds they hoped would use it. Rather, the prospective customers who were the designers' target probably were people who would make their choice based on what they wanted hummingbirds to be. After all, as Julius Caesar used to say, "People find it easy to believe what they desire to be true."

Is it possible to get an understanding of humming birds that is based on observation rather than desire, and which is consistent, at least potentially, with further scientific investigation? It seems clear that at least some types of comparisons between these birds and other creatures do not have any scientific sense or interest. For example, my newspaper once carried a short article under the title "Mondo hummingbird," reprinted from The Allentown (Pa.) Morning Call, in which Arlene Koch, a self-taught bird expert, described hummingbirds as rotten, nasty little creatures that only care about themselves. Furthermore, Ms. Koch was reported to say, after male and female hummingbirds travel thousands of miles from their winter havens in the south, they search out each other during the spring, whistling to each other at a level so low that humans can barely perceive it. When a pair hooks up, sex lasts only a few seconds. Afterward, the male goes out to find another female, without trying to take responsibility for his offspring. Once baby hummingbirds are just a couple of weeks old, their mothers kick them out of the nest. In spite of all this, however, Ms. Koch assured her readers that she loved those birds, as proved by the fact that, every year on April 15, she regularly set out nearly two dozen backyard feeders for them. (*The Globe and Mail*, Monday, March 20, 2006, p.A14.)

I do not believe it makes literal sense to describe hummingbirds as nasty, the way Ms. Koch did, since they are not humans. It might be appropriate to say that, if they were humans, then they would deserve to be called rotten and nasty; but that's a different story. In particular, hummingbirds, unlike us, are not creatures whose reproductive success depends on their engaging in cooperative behavior with other members of their species, or on having stable and long lasting familial relationships. Something else that makes it misleading for Ms. Koch to think of them in human terms, is the time scale in which they operate, which is much faster than ours. For instance, it is pointless to describe the sexes' few seconds of mating as stingy, unfeeling, or exploitative, since this sort of hummingbird behavior is normal, healthy, and adequate for its biological purpose, and therefore, in this sense, counts as good. Similar remarks apply to the two weeks of maternal care hummingbird chicks receive until they are able to fly, feed, and take care of themselves. Is it correct and sensible to describe those weeks as too brief? The right answer is: "Of course not!"

However, it need not be scientifically misleading to compare one species with another in all respects and cases. For example, one can obtain a fairly accurate idea of how hummingbirds survive, reproduce, and make their living, by taking account of their comparatively distinctive properties, organs, and behaviors—e.g. their small size, rapid flight, whirring wings, ability to hover, their long beaks and tongues, and the dates and routes of their annual migrations. Furthermore, it also is possible to draw inferences from those features, about the historical, evolutionary, and ecological relationships in which hummingbirds stand to their surroundings, including other living things. For example, evolutionary biologists tell us that that those birds are closely related to wrens, since both they and wrens evolved from a relatively recent common ancestor. Nevertheless, hummingbirds neither look like nor act like wrens today. Presumably, therefore, there was a time at which some of their forebears (they developed in the New World, not the Old) left behind a style of life that was similar to that of present-day wrens, and moved into the new, apparently unbird-like ecological niche they now occupy, which later proved to be one that was extremely rich, stable, and effective. Before the time of that change, only certain insects—e.g. bees—had occupied a niche that was analogous to the one hummingbirds now inhabit.

I find this last point about the history of hummingbirds fascinating, because we seem entitled to surmise that something similar must have happened in the evolutionary development of our own species as well. In other words, analogous to hummingbirds, it is arguably the case that humans also owe most of their recent biological and reproductive success to the fact that some of their evolutionary forebears stumbled upon an unexpectedly fruitful means of surviving, which happened to be vacant at the time it was needed.

However, there is a problem that infects any attempt to use those birds as a pattern, guide, or model for gaining an understanding of the place that our own species occupies in the world today. The problem to which I refer is that it is much easier to discover how hummingbirds are related to the things around them, than to understand how humans—both those now alive, and those that preceded them in time—do and once did interact with their environment. For example, humans have an unusual combination of bodily features, which unites adaptive styles of life that are characteristic of creatures of quite different sorts. The biologist J.B.S. Haldane was fond of pointing out (as noted by William Calvin, 2002, p.86), that only a human can swim a mile, then run 20 miles, then climb a tree. Swimming a mile is something deer are able to do, but wolves and chimpanzees cannot. Running 20 miles is within the capability of both deer and wolves, but not of chimpanzees. And neither deer nor wolves are able to climb a tree, but chimpanzees can. A similar point also applies to various "chosen" behaviors in which humans engage. Thus, behavioral ecologists often distinguish between what they call "r" and "K" reproductive strategies. The first strategy is followed by animals like mice and rabbits, which breed early and often, and thereby produce large numbers of offspring that, on average, suffer high mortality. The second is followed by animals like elephants and whales, which breed later and comparatively rarely, and thus produce fewer offspring, with lower rates of mortality, because the parents (or at any rate, one of the parents) invests a great deal of time, effort, protection, and care in the raising of each of those offspring. Which of these strategies does our own species follow? The answer is that humans—unlike the great majority of other animals—practice either the one or the other, depending on the particular circumstances in which they happen to find themselves. To be more specific, relatively impoverished humans who live without many of the benefits of modern technology tend to be "r" in their reproductive style. But if and when their socioeconomic conditions improve, they nearly always adopt the "strategy K" in place of "r," so that their infant mortality plummets, along with the average rate at which they produce children.

Analogous to what I said before about hummingbirds and wrens, it is not a promising explanatory strategy to try to predict how members of our species will behave by comparing them with their closest living primate relatives—the common chimpanzees and bonobos of Africa. To consider a simple instance, with only a few exceptions, when young common chimpanzee females come to be of breeding age, they usually leave the group in which they were born, and go to live with some other group of chimpanzees. But bonobos follow the opposite pattern. That is, when young bonobo males reach the age of sexual maturity, they are almost always the ones that leave the group in which they were born, and go to live out the rest of their lives in another group of bonobos. In the case of humans, however, neither the first pattern, nor the second, is followed. Instead, humans engage in a helter-skelter of many such arrangements, in which sexually mature young females sometimes leave the family group in which they were raised; sometimes young males leave their familial group; sometimes both males and females leave; and sometimes neither sex leaves.

Over the centuries, many attempts have been made to define humans in a way that would distinguish them from all other species. For example, some theorists have proposed that humans are the only animals that can speak; others say they are the only animals that make and use tools. I recently read about still another proposal of that sort that fits in with more contemporary concerns (see *The Globe and Mail*, April 8, 2010, p.L6)—namely, the idea that humans are the only animals that intentionally manipulate energy. I think that all the "definitions" of the kind just mentioned are too superficial to help me accomplish what I am trying to do in this book, because they refer to things that humans happen to do for many and various reasons, rather than focusing on the fundamental underlying powers and abilities that allow, enable, encourage, and sometimes even determine humans to behave as they do.

Most experts agree that the most distinctive organs humans have are those that make up their central nervous systems, prominently including their brains. For example, we now are aware of the facts (as the ancient Egyptians never managed to do) that the human brain is the source and focus of people's personalities, and also that the brain accomplishes many thousands and perhaps even millions of important functions, the great majority of which are quite different from all the others. Furthermore, there is no neat way of summarizing those functions, or of tying all of them together. The usual practice today of talking about the human brain as if it were a single organ is misleading, since it is more accurate to describe the brain as a very large group of different organs, whose only common feature is that they happen to be located in the same general place—namely, at the top of the spinal column and inside the skull. For these reasons, then, it is a difficult problem to find an accurate way of characterizing the "human niche" which, as far as we know, had no occupants on this planet, before some of our ancestors found it and moved into it. Another, related aspect of that same problem is that we also have no quick and easy means of comprehending the "human nature" that presumably has enabled the members of our species to survive in that niche.

(A digression: Someone might say it is easy for makers and consumers of science-fiction books and films to imagine other, non-human life forms, living both inside and outside of our solar system, galaxy, and universe, who or that inhabit essentially the same niche as we humans do. My reply to this objection is that it only poses an apparent rather than real difficulty, because the seemingly strange aliens that appear in movies and books do not amount to much more than duplicate humans; and because of that, they merely reintroduce all the old, familiar problems that are connected with attempts to understand and describe human nature, but fail to suggest any genuinely non-derived contrasts that might be of use for giving us instructive perspectives on our own situation. For example, no matter how seemingly exotic the shapes, movements, and sounds of imagined extra-terrestrial heroes and monsters might seem to be, my impression is that those in English films consistently think, act, infer, judge, and emote like Englishmen, those in American films like Americans, those in Japanese films like Japanese, and so forth. Thus, if someone really is interested in unraveling the mystery of what human nature is, I suggest that he or she could make more profitable use of time if, instead of paying attention to science-fiction, he informed himself about empirical discoveries that already have been made about the non-human intelligences that belong to other known terrestrial species. For instance, investigators could consider the intellectually influenced behavior of crows, parrots, and whales or—

perhaps even better—of cephalopods like octopi, nautiluses, and squids.)

Even though I teach philosophy for a living, I also have been interested for a long time in the topics of humans' political, military, artistic, and social history and prehistory. Perhaps because of that, I gradually have formed the opinion that quite a few important but relatively neglected parts of present-day human nature have come into existence as a result of cultural advances made by some of our ancestors. In other words (in contrast with what most of my philosophical and scientific colleagues believe), I do not think it is possible to make sense of what present-day humans are like just by paying attention to details of our physical environment—prominently including the story of how our bodies are composed and organized—because some aspects of human nature are not just products of biological evolution, maturation, health, and growth, but are, at least in part, expressions of past cultural innovations (i.e. thoughts, discoveries, ideas, and choices) that were thought of and brought into existence by now unknown, long-dead individuals.

1.2 Is there any such thing as human nature?

An English proverb runs as follows: "The fox knows many tricks, the hedgehog only one; but it is a very good trick." Some people interpret this proverb as implying that it is right to speak of hedgehogs as having a nature, since a nature is an unlearned, inherited, and instinctively automatic program of behavior, and the hedgehog's trick of responding to every perceived attack by rolling into a ball that puts its soft, vulnerable parts on the inside, and its stiff, prickly spines on the outside, is behavior of exactly that sort. On the other hand, however, similar reasoning also implies that foxes do not have any nature since (1) depending on the characteristics of each situation, foxes act in a far great number of possible ways than hedgehogs do. And still more specifically, (2) foxes apparently have a power to choose among their behaviors, by appealing to certain needs and desires as a means of deciding what to do in each case.

Some philosophers argue on grounds analogous to what I just said about hedgehogs and foxes that there also is nothing that correctly deserves to be called "human nature." For example, the French atheistic existentialist philosopher, Jean-Paul Sartre (see his 1948/1964, pp.122ff.), claimed that animals like streptococci, starfish, rats, hedgehogs, eagles, squids, kangaroos, and even foxes, all have a settled and more or less predictable nature that scientists can discover and describe by means of observation and inference. But the same was not true of human beings, according to him, because each human was able, at every moment of his or her life, to transform himself (or at least to begin to do that), in any direction, and to any degree, he chose. (Of course, this idea also presupposed that each of the humans in question was willing to be sufficiently determined, brave, and mentally strong to bring about the changes he or she desired.) To summarize, then, Sartre's view was that humans did not have a nature, because they had absolute freedom, and freedom of that sort was incompatible with their having a nature.

Another means Sartre sometimes used to express roughly the same idea was to say there always had to be a plan, blueprint, or essence that preceded the bringing into existence of any manufactured object like a letter opener, automobile, house, or sausage factory, since the first step any would-be manufacturer would have to take in order to produce the item in question, and thereby make it real, would be to consult a plan of that sort, which would allow him or her to envisage all the proposed object's properties. However, according to Sartre, there could not be any such plan in the case of human beings, because human behavior (unlike that of all other creatures) was not determined either by natural or by supernatural influences. The proof for the first part of this idea—about natural forces—lay in the perceived fact of complete human freedom. (Sartre described this freedom as "something we all feel.") The proof of the second part—about supernatural forces—lay in the further, metaphysical fact (presumably established by means of scientific and historical observations and reasoning) that there was no God, who might bring into effect a program of first envisioning, then also creating humans. Accordingly, then, (i) it was necessary for human beings constantly to remake themselves, without having any pre-existing conception to work from; and (ii) the only thing that determined what they were like at any moment of their lives was this self-envisaging and self-creative process itself. Sartre then added the further, still more general point that (iii) human beings were unique among living creatures, and perhaps among all objects considered in general, because of the fact that their existence preceded their essence.

Let me begin to criticize Sartre's conception by saying that his idea that humans have no instincts strikes me as being suspiciously unscientific, because it fails to take account of many clear facts and cases. For example, several streams of linguistic research point to the conclusion that humans would not be able to learn any language of any sort, without the help of a very large number of innately given and automatically triggered thought and behavioral routines.[1] Furthermore, as opposed to the Sartrean idea that humans are free to learn any language they like, linguists' observations repeatedly have confirmed that there are many logically possible languages that humans cannot learn. The most plausible explanation of why people cannot do this is that their inborn instincts impose severe limitations on their abilities. The following provides a simple instance of this point. Speakers of English find it easy to form questions from declarative statements, by transposing the main verb to the beginning of the statement, and leaving an unspoken and unheard "trace" in the place where that verb used to be. Thus, we generate the question corresponding to the descriptive statement, "Your horse is brown," by changing the order of the words in that statement to say instead, "Is your horse [is] brown?" But our brains' organization does not allow us—except, possibly, in the case of a small number of very unusual people like Wolfgang Amadeus Mozart—to form questions by the equally simple, but systematically different, method of reversing the order of the words in the original sentence. That is, the vast majority of humans cannot form questions from a sentence like "Your horse is brown," by saying "Brown is horse your?" (Again see Pinker, 1994, Chapter 4.)

Turning now to a theory of another type, there are some thinkers—for example, the behaviorist psychologist B.F. Skinner—who admit that humans have a nature, but then deny that this nature applies to our species alone. To be more explicit, the people about whom I now am speaking believe that all comparatively complex, conscious, and changeable creatures, including humans, determine their patterns of thinking, of behaving, and of gathering new information, in essentially the same fashion. Furthermore, they claim that every living thing that is capable of learning anything at all, learns only those things that lead to reward (or what some behaviorists call "positive reinforcement") and either ignores, or explicitly avoids,

---

[1] It seems to me that Darwin in his book, *Descent of Man,* William James following him, and Noam Chomsky following those other two people in turn, almost certainly were right to say that, rather than humans' having no inborn instincts at all, they have a far greater number of instincts than animals of any other known type. (On this point, see Pinker, 1994, pp.20-1.)

things that lead to a lack of reinforcement or to negative reinforcement. This implies in turn that all conscious organisms share, to one or another degree or level, in the very same type of character, mind, or nature. Presumably, behaviorists assume that humans have a more developed and complete version of the common nature or "group mind" about which I now am speaking, and that simpler animals have less complete versions of it. For instance, some biologists deny that very simple creatures like bacteria, ants, mosquitoes, etc. have a nature at all, on the grounds that they are automatons that never learn anything. To be more specific, even though it often seems to us that creatures of that sort are behaving in ways that are considered, useful, and adaptive, experimental investigations have shown that everything they do is a result of programs of behavior that are rigidly determined by their genes, so that what they do is not actually a result of deliberations or choices. On this subject, see Wilson 1978/2004, p.55-6.

I believe the truth about the matters now under discussion falls somewhere between the two extremes just mentioned. Sartre's idea about human beings' having absolute freedom strikes me as an unrealistic exaggeration, inspired by nothing more than wishful thinking and a Marxist political ideology. It occasionally may be useful to affirm "the truth" of some such idea to children and other naïve and inexperienced people, as a rhetorical means of encouraging them to be more creative, ambitious, imaginative, and courageous. But this point is not nearly sufficient to establish the literal correctness of this way of describing all human beings. Rather, both scientific research and everyday life experiences indicate that humans like us are partly free and partly not free. In other words, it is clear from repeated observations that there are respects and situations in which we can change some of our values, habits, behavior, and attitudes, in certain directions and to some extent, just by deciding to do so; but there also are a vast number of other cases in which we cannot do that.

Correspondingly, I also do not think Skinner is right to claim that simple rules of learning determine all of the knowledge and behavior of conscious and adaptable creatures, so that all creatures of that sort share to one or another degree in a single, overarching mind or nature. One indication of the lack of justification of this claim is the observed fact that each species of animals is "prepared" to learn some stimuli, is barred from learning others, and is neutral (i.e. has some sort of meaningful choice) with respect to still others. As illustrations of this point, consider three cases E.O. Wilson mentions in his book, *On Human Nature* (1978/2004, p.65). First, Wilson tells us that adult herring gulls quickly learn to recognize and distinguish their newly hatched chicks by sight; but they never learn to recognize their own eggs, even though the eggs are just as visually distinct as the chicks. Second, he says that in spite of the fact that newborn kittens are blind, unable to walk, and nearly helpless, they possess an advanced ability to learn, in respect of the relatively few matters on which their survival directly depends. For example, in less than one day, kittens learn, by smell, to crawl to the spot where they can expect to find the nursing mother; and each one also learns by the same means how to find its preferred nipple. Third, Wilson talks about birds known as indigo buntings that migrate at night from their breeding grounds in eastern North America to their wintering grounds in South America, by taking their bearings from the stars. Tests have established that those birds quickly learn the look of the circumpolar constellations (the ones that surround the North Star), but that they are inhibited from learning any other constellations.

If human beings really were free to learn whatever they wanted and needed to learn, then it might have been correct for Sartre to say we have no nature. Correspondingly, if Skinner and other behaviorists were justified in supposing that humans only learn and thereby know things for which they are positively reinforced, then although humans would have a nature, it would not be a distinctive one, since other conscious creatures would share in that very same nature. My comment on all this is to propose a middle hypothesis—namely, that humans' powers of learning are selective and limited, since (a) some of their knowledge is given to them innately, (b) other bits of knowledge are impossible for them to acquire, and (c) they learn still other bits of information only if and when they choose and set out to learn them. Furthermore, if this middle hypothesis is correct, then it is appropriate to say that human nature is something that at least is relatively distinctive after all. Nothing said so far in this chapter implies that humans either do have or must have exactly the same set of inborn instincts as all other conscious animals. Furthermore, the preceding points also do not imply that every human being has exactly the same instincts as every other human. For instance, some of us are right-handed, others left-handed, and still others are ambidextrous; some of us are able to roll r's, while others of us cannot do this. Nevertheless, I believe Wilson is right to say there is a sufficiently extensive and powerful convergence among the innate behaviors that belong to humans to justify the idea that we do, after all, have a shared nature that is distinctive for our species. (Again, see Wilson 1978/2004, p.67)

Let me now end this section with a short summary. It is a datum of experience for us that humans, along with some other animals like foxes, have a certain amount of freedom of choice, as shown by the fact that we can deliberate and choose on the basis of our deliberations, between alternative courses of action. However, neither Sartre's theory of human nature, nor Skinner's, succeeds in doing justice to that observation, because the first exaggerates the extent to which we are free, and the second underestimates it. Furthermore, another mistake implicit in the second, Skinnerian theory, in my opinion, is that it underestimates the amount of behavioral distinctiveness that belongs to humans, as compared with the behavior of other creatures.

1.3 Did a "Linguistic Rubicon" occur at a certain point in human history?

At first sight, present-day theorists seem to have a great advantage over people who speculated about the topic of human nature in the past. The reason for this is that observational and experimental science has provided recent investigators with empirical data and other relevant resources that are richer, more detailed, and better confirmed than those that were available (say) to the authors of the Upanishads, to Confucius, to Heraclitus, to Aristotle, to Averroës, to Thomas Hobbes, and to Charles Darwin. However, at least as far as concerns fundamental problems of the kind we are trying to solve in this book, I believe that advantage is more apparent than real.

A case that shows what it means to say this is the following. Can we draw a significant distinction between modern human beings on one side and all the other creatures that ever lived on the other? In spite of the many intellectual advantages available to us today, there still is widespread disagreement about this question at the present time, which divides one group of researchers from another. Some of the people to whom I now refer (e.g., Noam Chomsky in his book 2000, especially p.4) take it to be obvious that human beings are different from all other creatures, because according to them it is an observable fact—shown, in particular, by our ability to speak in a way that is not available to animals of any other kind—that humans represent a radically new theme in history. In particular, Chomsky says in the book just mentioned that human language could not have evolved in a gradual and orderly way from the signs, gestures, and vocalizations of other organisms, but instead must have come into existence as a result of some special, unnatural, and lucky "jump," because there are and were no other pre-linguistic organisms from which language might have evolved in a Darwinian fashion. By contrast, other people (e.g. Desmond Morris, 1967; Jared Diamond, 1992/2006; and Barbara J. King, 2001, 2002) continue to insist that (i) Darwinian natural selection is the only principle capable of providing a legitimately scientific account of how organisms change, and (ii) natural selection presupposes that all living beings are linked together in a world-sized web of historical relations. Thus, members of this second group conclude that (iii) humans' seemingly unique ability to speak does not show that they are significantly separated from all other organisms, because if one looks closely enough, one will be bound to see that quite a few of our fellow creatures have powers and capabilities that at least are proto-linguistic in character. (As an example of this kind of approach, consider the title the biological anthropologist Barbara J. King gave to her first published book—namely, *The Information Continuum: Social Information Transfer in Monkeys, Apes, and Hominids*.)

In the year 49 B.C., there was a sudden, serious, and irreversible change in the political and military career of the Roman general and politician Julius Caesar. This happened when he decided to lead his army across the Rubicon River from Gaul into Italy, in defiance of a Senate decree that explicitly had forbidden him to do so. In a partly similar style, investigators of human nature now ask whether, at the time certain members of our species finally acquired the use of fully-fledged, syntactical language, this also had the effect of quickly, decisively, and irreversibly transforming them into creatures of a different sort than they had been before. Still more specifically, people who debate this question ask whether the hypothetical event just mentioned—a "linguistic rubicon"—(assuming, for the moment, that it actually did happen) changed our ancestors in such way as to make them, for the first time, recognizably similar to ourselves.

In my view, the debate just described is misconceived. It does not make sense for people to engage in a dispute about whether or not a linguistic rubicon occurred at some time in our species' history, because a great many archeological observations already have established it as an undeniable fact that an event something like that—indeed, a whole cluster of them—really did take place. Those events happened at different times, and in separate locations in the Old World, between the dates of approximately 60,000 to 30,000 years before the present. Furthermore, this group of similar events had such a significant effect on the lives of our ancestors that scholars now accept those occurrences as marking the beginning of a whole new phase of human pre-history—namely, the "upper" or later phase of the Paleolithic era (or Old Stone Age). Accordingly, therefore, the name informed people now bestow on these happenings is the Upper Paleolithic Revolution (to which I shall refer hereafter as the UPR).

What happened at the time of the UPR? To answer that question, let me refer to a passage from a paper written by the anthropologist Paul Mellars, in which he lists seven typical changes that occurred at the start of the Upper Paleolithic era, each one of them extensively documented from archeological materials (1998, p.92).[2] The first point he mentions is the appearance of much more widespread "blade" and "bladelet" as opposed to "flake"-based technologies. The second thing is the appearance of a wide range of entirely new forms of stone tools, some of which reflected an entirely new level of visual form and standardization in the production of those tools. The third point to which Mellars refers is what he calls an "explosion" of bone, antler and ivory technology, that involved a remarkably wide range of new and tightly standardized tool forms. The fourth point is the appearance of the first reliably documented beads, pendants and other items of personal decoration. The fifth item is the transportation of seashells and other materials over remarkable distances. The sixth is the appearance of the first unmistakable sound-producing instruments. Finally, his seventh point is the dramatically sudden appearance of explicitly "artistic" activity, in a remarkable variety of forms.

(Mellars fails to mention at least two important additional points. These are (a) that this was the time when people first invented ocean-going boats, which played a role—almost immediately after the very first known instance of the UPR—in the human colonization of the then combined lands of New Guinea and Australia. Furthermore, (b) he also says nothing about the fact that the UPR was followed by a dramatic rise in the numerical world population of human beings of our species.)

What connection is or was there between language on one side and the UPR on the other? The simple answer is that anthropologists

---

[2] In view of what I take to be the importance of this passage, I have quoted it in two of my previous publications as well—namely, in 2003, p.15, and in 2005, pp.505-6.

assume there must have been a very important and intimate link between these two things, because they cannot think of any other plausible explanation of why all the various developments mentioned before should have occurred at roughly the same time, except to say that they happened (i) because this was when humans learned to speak in terms of full-blooded, syntactical language, and (ii) their having acquired a language of that sort allowed them to think in ways that were much more powerful and innovative than the kind of thinking they had employed before.

I grant the truth of the claim that the UPR is the most obvious, dramatic, and most thoroughly scientifically documented of all the intellectual changes through which our ancestors passed in the prehistoric era. Nevertheless, that admission does not imply that the UPR also was the earliest, or the most basic, or the most important of all the changes that took place in approximately the same period. Similarly, even if it is true to say that the first appearance of whole or "finished" syntactical language took the form of a comparatively sudden, rubicon event, that still does not justify the idea that language was (as far as we know) something that was new and separate from anything else that previously had existed. Rather it still is possible, analogous to what people like Barbara J. King say, that before the time of the UPR, language developed in a gradual, incremental, and progressive manner from a primitive, still "unfinished" state into something that our forebears suddenly became able to employ as a means of thinking in terms that were general, abstract, and explicitly symbolic.

What sorts of things might have preceded and prepared the way for the appearance of syntactical language and the UPR around 60,000 to 30,000 years ago? Although I talked about the UPR in my earlier authored book published in 2003, I did not discuss this topic of preparation there. Nevertheless, I do not subscribe to the idea that syntactical language somehow appeared in a mysterious, miraculous, and unexplainable fashion. Thus, in the light of this point, one of my main projects in this second authored book is the previously omitted one of trying to take some of the mystery out of the sudden appearance, in the general history of life, of syntactical language.

1.4 The thesis of this book: The distinctiveness of present-day humans (including separateness stemming from their use of language) is not just a passive product of biological and historical changes, but also was partly self-created

> It's frightening to think that you mark your children merely by being yourself.
>
> Simone de Beauvoir

In a cover article about the Middle Awash area of Ethiopia that recently appeared in the *National Geographic* magazine (July, 2010), author Jamie Shreeve described this part of East Africa as the most persistently occupied place on Earth. What he meant by this remark is that members of our evolutionary linage have lived, died, and become buried there through a time span of almost six million years. At the present time, furthermore, the bones of those ancestors are eroding out of the ground. Thus, Shreeve asked in rhetorical fashion, what better place could there possibly be to help us learn about the circumstances, conditions, and changes that eventually made us into humans?

I certainly agree that we can learn a great deal about what allowed us to develop into the kind of human beings we are now, by observing and analyzing the bones, teeth, stone tools, living sites, and other artifacts associated with the long line of our forebears, which paleoarcheologists have discovered—and will continue to discover—in the soil and rock of places like Ethiopia's Middle Awash area. Nevertheless, my main reason for having written this book is that I do not believe considerations like the ones just mentioned were the only sort of factors that led to our becoming what we now are.

In the book Charles Darwin published in 1859, he argued that living creatures changed and evolved through the workings of two closely related mechanisms or principles, to which he gave the names of natural selection and adaptation to one's environment. According to him, those principles, working together, brought it about that the particular characteristics and behaviors of any organism, which happened to be useful to it in its struggles to survive and reproduce itself, were more likely than its other characteristics to be preserved by being passed down to that organism's descendents. However, neo-Darwinist critics like Stephen Jay Gould have pointed out that, in addition to those two basic notions, factors of quite a few other sorts also have played important roles in determining which organisms survive, and what properties those organisms have.

What are the additional factors about which I am talking? Consider a simple example. Occasional catastrophic events like volcanic eruptions, earthquakes, tsunamis, and asteroid hits have killed a great many creatures, and also have killed them in a more or less selective manner. Because of that, events of that type have had a profound effect on the types of organisms that continue to inhabit our world today, and on the properties those organisms now possess. Nevertheless, the means by which such occurrences influenced the course of life on Earth were not just by their providing additional contexts and occasions in which ordinary Darwinian natural selection could do its work. A proof of this last point is to note that—from the viewpoint of the individual creatures involved—whether or not they survived any given catastrophe was only a matter of luck, rather than a matter of their becoming adapted to the new environment that the event in question had created. The reasons for this in turn are that (i) the creatures affected did not have any means of foreseeing, predicting, or preparing for the catastrophic occurrence; and (ii) the vast majority of the organisms involved also did not have sufficient time to adjust themselves to the new circumstances associated with that event.

I think Gould and other neo-Darwinians are justified in expanding the scope of Darwin's original vision in the way illustrated by the preceding example. Nevertheless, it also seems to me that those critics did not expand Darwin's view far enough. What do I mean by this? What I mean is that ordinary experience makes it clear that people of our sort are constantly being influenced, not just by things we see, hear, and feel around us, but also by items like thoughts, conceptions, and ideas—either thoughts that we ourselves have brought into being, or ones that other people have suggested to us. Against the background of that point, the thesis for which I shall argue in the rest of the book is that the human nature we now possess is partly based on a series of intellectual inventions made by various, now unknown individuals who lived in prehistoric times.

Many modern researchers (including some neo-Darwinians) take the findings of natural and physical sciences like physics, chemistry, biology, physiology, and anatomy much more seriously than discoveries made by practitioners of social and historical sciences like archeology, anthropology, and history. Perhaps because of that, those critics have failed to see that—at least as far as our own species is concerned—biological and physical conditions have not been sufficient to determine what we have become, since there have been important contributions from thought and culture as well. To say the same thing another way, not all of the factors that produced present-day human nature were of the sort that investigators might have dug out of the ground. Accordingly, rather than simply finding and observing the cultural factors that influenced our development, we are obliged to discover them by means of reconstructive inferential reasoning. In other words, the cultural factors about which I now am speaking are (and were in the past) internal to humans rather than external. They also are active rather than passive, in the sense of being items that people had to think of, accept, and choose. In this respect, they were quite different from developmental factors that people did not and could not choose, but by which they were changed, either against—or at least irrespective of—their wills.

Finally, one more thing I want to do in this introductory chapter is to set forth a simple, clear, and hopefully not overly controversial, parallel case, adopted from recent sociological investigations, which illustrates what it means to say that cultural traditions are able to change some of the ways people habitually think and act. What accounts for the fact, repeatedly confirmed by economists, historians, and psychologists, that people born into the Ibo tribe of Nigeria tend to be more economically and professionally successful than members of the other tribes of that country? It strikes me as a fairly obvious point that natural selection would not and could not have made the members of that particular tribe more intelligent, organized, and ambitious than other Nigerians. Rather, it is more plausible to believe the crucial factor in this case is a cultural tradition that the Ibos themselves inaugurated a long time ago. In particular, one central part of this tradition is that, in the funeral rites and ceremonies held for a deceased Ibo, all the honors, titles, offices, and most of the wealth that once belonged to the dead man or woman are laid into the grave symbolically, along with the corpse. This serves as a reminder to both the children of the deceased, and to all of the people who know them, that none of those things is available to be inherited by the children, and therefore whatever those children achieve in their lives, they will have to earn through their own, unaided efforts.

The individuals who started this tradition did not "plan" for their descendents to compete more successfully for money, jobs, influence, and prestige with representatives of Western and Westernized societies in Europe, North and South America, Japan, China, India, etc. Rather, those descendents now find themselves in a position to compete more effectively than other people from their country, as a result of nothing more (from their point of view) than blind circumstances and luck. Nevertheless, the cultural tradition in question does have this kind of adaptive effect, just as surely as if that result had been produced by natural selection.

Is it right to suppose that the human nature that belongs to nearly all present-day people is relevantly similar to what I just said about the Ibo tribe, because it too was partly formed by past cultural innovations? I believe the answer is yes; and my means of arguing for that conclusion is to point out that we have good archeological reasons for believing there once existed, in pre-historical times, a small group of "founding" cultural traditions, whose influence extends up until the present moment. To be still more specific, one thing those founding groups handed down to their successors was a habit of thinking (probably invented independently several times, by each of a relatively small group of people) of our species as having been set apart from everything else in the world, by something like divine intervention. I claim that this is similar to the attitude of the Ibos described before, because it also has given modern humans an advantage over all the other species on our planet, which natural selection, just by itself, could not have made available to them.

Chapter 2

Darwin and his Successors have Not Taken Proper Account of Culturally Created Human Characteristics

2.1 What modern humans are like: A tangled knot science only has begun to untie

The more unintelligent a man is, the less mysterious existence seems to him.
Arthur Schopenhauer

My attitude towards Charles Darwin is one that is respectful[1] and even, in some ways, protective. For instance, I agree with the statement made by the biologist G.G. Simpson (quoted by Richard Dawkins at the beginning of his book, 1978) that, since the dawn of history, many reflective people have asked questions like, "What is the meaning of life?," "What exactly are human beings?" and "Why do we exist?" But none of those answers had any genuine or lasting value—until the year 1859, when Darwin published *On the Origin of Species*. What did Simpson think was the historically decisive accomplishment Darwin's book had made? Part of the answer is that he supposed that practically all of our ancestors—inspired by religion, romantic idealism, feelings of self-importance, etc.—conceived of humans as having some special status and place relative to the universe as a whole. But Darwin's revolutionary insight was to show that this popular idea was false, because it was not only possible, but probably also was true, that instead of being products of the plans and purposes of an all-seeing God or group of gods, we and all other creatures were nothing more than outcomes of certain blind and unreasoning natural processes. Thus, Darwin's book forced us to begin the hard work of piecing together an alternative picture of human life and nature that could be consistent with the empirically grounded discoveries that recent scientists have made (including Darwin himself, but not limited to him).

---

[1] For example, I said in the preface to my previous book (2003, p.x) that I accepted Darwin's style of thinking, reasoning, and writing as a model for my own work.

Nevertheless, I do not think of Darwin's views as an infallible gospel. Let me give an example to show what this means. Several years ago I noticed a vehicle in one of the parking lots at my university, which had a cross hanging from its front, rear-view mirror, and a raised, metal decal on the back of the vehicle, that was roughly in the shape of a fish. As I got out of my car and prepared to walk toward my office, I took a closer look at the decal. It was a picture of a large fish, on whose side was written "Truth," which was in the process of swallowing a smaller fish, on which appeared the name (or part of the name) "Darwin." This gave me a very uncomfortable feeling. I just then was on my way to teach my weekly class on philosophy of biology, and was carrying the course's kit of reprinted articles in my briefcase. The first of those articles, with which the course had begun one or two weeks ago, was a piece by David Quammen from the *National Geographic* magazine, reviewing the main lines of evidence—biogeography, paleontology, embryology, and morphology—to which Darwin had appealed in *On the Origin of Species*, in defense of the principle of natural selection. Quammen's article carried the title, "Was Darwin Wrong?" (2004); and one feature of that piece that struck me as a clever way of saying something true and important, was the fact that its first word, printed in very large capital letters, was NO.

When I first saw the decal, something like the following thought came into my mind. "If universities are supposed to be places where people discover, confirm, and celebrate truth, and expose and reject errors and falsehoods, then 'WHAT IS THIS CAR DOING IN A UNIVERSITY PARKING LOT?!'"

(Let me add that I am not personally opposed either to religion considered in general, or to Christianity more particularly. For example, I regularly teach a university class on philosophy of religion, and I attend church services with members of my family on most Sunday mornings. Nevertheless, biblical literalism strikes me as an extremely silly business that any person who takes the time to study the Bible in even a moderately serious fashion soon ought to abandon. Thus, although the cross hanging from the car's mirror was not offensive to me, the same was not true of the decal.)

Since then, however, the experience of continuing to teach courses in the area of philosophy of biology has led me to see that there are quite a few respects in which the decal was correct after all. Thus, I now feel a certain amount of shame about the smugly intolerant thought that came into my mind when I first looked at the car's decal, because I now consider that thought to have been a contribution to an idea recently expressed by some newspaper columnists and editorial writers (see e.g. the column of Margaret Wente, *The Globe and Mail*, June 17, 2006, p.A21, and one of that newspaper's editorials on the same day, p.A20) that some Canadian and American universities have become "... islands of repression in a sea of freedom." Still more explicitly, it does not seem to me that Darwin was wrong relative to the claim of biblical literalists that God created the world and all its creatures over the span of six days, but I do think that what most people have in mind whenever they talk about "Darwinism" is confused and positively mistaken in at least two other respects.

The first of those respects is not a very controversial one. In fact, it is something Darwin himself recognized and accepted, although his colleague, Alfred Russel Wallace, the co-discoverer of the principle of natural selection, did not. As Steven Jay Gould said in one of his papers (1980b), Wallace subscribed to a very strict form of Darwinism, according to which natural selection was supposed to be able to account for virtually every property, part, and organ that belonged to every creature; but Darwin did not subscribe to that same idea. Rather, Darwin believed (and said explicitly) that natural selection was just one, but not the only, means by which organisms developed and changed. Since Gould thought Darwin was right to take this position, and Wallace was wrong, he followed the following "two-pronged" approach in expressing his own relation to Darwinism. On one side, he said he accepted Darwin's principle of natural selection as basically correct, and of great scientific value. But on the other side, he also recognized that there were many well-established facts about how organisms changed and acquired their properties, which the principle of natural selection, narrowly conceived, still was not able to explain. Thus, Gould believed Darwin's theory needed to be supplemented in such a way as to allow it to take account of additional truths, incompatible with the principle of natural selection, which more recent investigators had uncovered. (On this subject, see Tim Flannery's summary of Gould's views, in 2002.)

Consider a pair of simple instances of such "additional truths." First, why do male humans, like the males of all other species of mammals, have nipples, even though those nipples are not able to supply milk to babies? The answer is that natural selection did not produce those organs in order to give men a competitive advantage in the struggle for existence; and therefore they do not count as adaptations. Instead, they are parts or aspects of a historically determined, common body-form in which both men and women share equally. In the case of males, their nipples have no direct biological function, but are merely "dragged along" with the male version of that same bodily form.[1] Second, it also is wrong to suppose that mammals displaced dinosaurs from ecological niches that were capable of supporting large carnivores and herbivores, because they had "beaten dinosaurs at their own game, by showing themselves to be more fit." Rather, Gould claimed that dinosaurs became extinct for reasons that were completely independent of the relations they had with our ancient, rat-sized, mammal ancestors. Those reasons included the strike of a large extraterrestrial object near or on the Yucatan peninsula, 65 million years ago, which caused a widespread extinction that helped to clear the way for mammals like us to move into those newly vacated niches at a later time.

However, there is also a second respect in which it seems to me that what Darwin said in his book was mistaken, which is not equally familiar to his critics. One possible reason critics have not paid much attention to this second difficulty is that Darwin, similar to many people who are alive at this present moment, was far more concerned to talk about the similarities and commonalities among living things than to talk about their differences. Nevertheless, (echoing the kind of position Noam Chomsky once defended) that attitude led Darwin to overlook an important respect in which humans really do count as different from all other creatures.

Darwin began his "one long argument" for the diversity of life and the mutability of species in *On the Origin of* Species, by engaging in a thorough discussion of what he called "artificial selection." (See Ernst Mayr, 1991, and the Introduction to Darwin's *The Descent of Man* by H. James Birx, especially p.x.) This last phrase referred to the long-established practice among farmers and breeders, of improving the health, productivity, and suitability of their domesticated plants and animals, by always choosing to breed those individuals that were best in the respects in which the farmers happened to be interested, and

[1] The medieval philosopher, St. Augustine, had a systematically different idea about the notion of "biological function" from the one Darwin employed. This conception led Augustine to speculate that God had given nipples to men in order to increase the beauty of their bodies. (See the translated passage from *City of God* XXII, 24, in Bourke, 1978, p.119.) However, I confess that Augustine's attempted explanation leaves me cold, because—constructed as I am—I am not able to see what it is about male nipples that prompted him to consider them beautiful.

not breeding—and/or killing—the individuals that were deficient in those same respects. After that introductory discussion, Darwin then switched to a consideration of what he called "natural selection," and argued in an analogical fashion that—in effect—nature also determined which creatures should live and reproduce, and which others should die without offspring. According to him, the most significant difference between that which nature did, and what farmers did, was that farmers assessed the "fitness" of their animals and plants in terms of certain advantages those creatures provided for the farmers themselves and their families; but nature only assessed the fitness of organisms in terms of the advantages that their characteristics provided for the survival and reproduction of those creatures themselves. (A third form of selection Darwin also discussed was "sexual selection."[2] This at least was partly similar to human controlled artificial selection, as shown by the fact that the criterion employed by females—including human females—to decide with which males they should breed, had very little to do with what was good for the males themselves, but had much to do with what was good for the females and their offspring.)

Now consider the following question. If what people usually understand by the word "choice" is making thoughtfully weighed decisions between separately distinguished alternatives, based on various values, purposes, and preferences of the chooser, then choice obviously is involved in farmers' artificial selection of plants, animals, and funguses. (Furthermore, it also might be involved in the case of human females who practice sexual selection.) But choice cannot also be present in the workings of natural selection, because (as many authors frequently have pointed out) natural selection operates in a way that is automatic, mindless, and value-free. In this sense, then, I suggest that the two things just mentioned are and must be fundamentally different.

How significant—and how important to the matters discussed in this book—is this difference? I believe that, strictly speaking, Darwin was wrong to conclude immediately from the legitimacy and effectiveness of artificial selection, to the presumed legitimacy of natural selection as well, because of the fact that the first item was not relevantly similar to the second one. (Note, by the way, that Gould's

---

[2] Sexual selection was a central topic of Darwin's second book (written 12 years after the publication of *On the Origin of Species by Means of Natural Selection*). The full title of that second book is *The Descent of Man; and Selection in Relation to Sex*.

characteristic method of defending Darwinism—i.e. by affirming the correctness of natural selection, but then "isolating" that principle from other means of determining organisms' properties—does not evade the force of this point, since the objection I am making here is that Darwin failed to provide good, clear, and defensible grounds for talking about natural "selection" in the first place.)

Some readers will feel that what I am saying here is both (a) unreasonably pedantic, and (b) contrary to obvious lessons of history. They will say that artificial selection and natural selection are sufficiently similar to one another, to justify the analogical method of reasoning Darwin employed, and to allow that reasoning to work, because Darwin's conclusion was based on a metaphor that was harmless. What does it mean to speak of a harmless metaphor? Let me provide a couple of simple examples. People often talk about certain ant colonies having slaves, and about all beehives having queens. But since ants and bees do not have either language or cultures that are inspired and organized in terms of language, they also do not have the concepts, traditions, rules, laws, etc. that are associated with, and partially constitutive of, the human institutions of enslavement and royalty. Thus it follows, in turn, that it cannot be literally correct to describe some of those insects as having slaves or as having or being queens. Nevertheless, in spite of that, reasonable people would say that the patterns of social arrangements in which those insects live are sufficiently similar to the structures of human slave-owning societies and human monarchies, to make the comparisons just mentioned worthwhile and informative. By parity of reasoning, then, even if it happens to be true in a strict sense that nature is not capable of making choices in the same reflective and considered manner that is characteristic of farmers and breeders, we still are justified in claiming that Darwin's comparison is a useful and enlightening one, because it helps us gain insight into various facts about the workings of nature that millions of people in past ages would have been extremely happy to obtain, if only they had been given an opportunity to learn about those facts.

It is easy to be sympathetic to the sanity and level-headedness of the reasoning just cited. Nevertheless, I continue to insist, contrary to the views affirmed by many other recent commentators, that Darwin's method of arguing in 1859 was not justified, for the following reason. Something that amounts to a harmless metaphor when people run it in one direction, may turn out to be a quite harmful source of error and confusion, if one tries to run it in the opposite direction instead. For example, I grant that it sometimes is clarifying to compare societies of bees and of some ants to human monarchies and arrangements between human slaves and slave owners, because of the obvious fact that bees and ants do not have any language and therefore also have no language-based cultural traditions. On the other hand, however, it does not seem to me that anything but confusion can be a result of following the opposite procedure of trying to explain why certain language-using humans have kings, queens, and slaves, by claiming that those institutions have arisen from the very same factors as those that influence bees and ants, because this willfully ignores the basic and crucial role that we know language and language-based thinking plays in human society.

Let me explain this same point in a slightly different fashion. Many theorists assume that if it is possible for us to learn various useful and significant things about non-humans animals, by using the Darwinian method of comparing natural happenings to some of the things that humans do, then it also ought to be possible for us to understand those same activities of human beings in a better and clearer fashion, by comparing them with the very same (kinds of) behaviors that are characteristic of non-humans, which I just mentioned. But this reasoning is fallacious because it does not really have the effect of illuminating what human beings are like. Still more specifically, I think the reason this reasoning fails to do this is that it denies, or obscures, or consciously sets aside something that we know is one of the most important and most distinctive characteristics of our species.

Consider an example. Is there an essential difference between (1) an adult male lion that kills and eats the cubs of a rival male he recently displaced from a leadership and mating role among a pride of lionesses (a clear instance of Darwinian natural selection), and (2) a man arranging clauses in his will so as to avoid leaving money to individuals to whom he is related only by law rather than by blood? Self-described sociobiologists like E.O. Wilson say the answer is no. But I suggest that this answer is misleading, because the man in our example may and probably does have reasons for doing what he does—no matter how shabby, flimsy, irrational, desperate, or otherwise questionable those reasons might prove to be when they are examined—but lions do not do what they do for reasons. How could someone discover precisely what sort of motivations lions have? My proposed answer to this last question is that scientific investigators need to design subtle and clever, but nevertheless solid, objective, and convincing experiments that are capable of throwing light on the question of exactly what does and does not lead male lions to kill cubs that were not sired by them.[3] In contrast with this, I believe it always is a mistake for would-be researchers to appeal to the Hobbes-like, anthropomorphic, "quick" method of simply trying to think themselves into the skin of one or another non-human creature, like (e.g.) a male lion that recently became the dominant individual in a pride.

There is a much-quoted passage on page 21 of Daniel Dennett's book *Darwin's Dangerous Idea* (1995), in which he says that if he were in a position to give an award for the single best idea that anyone ever had, he would give it to Darwin, ahead of Newton and Einstein and every other person in history, because of the fact that, in a single stroke, the idea of evolution by natural selection unified the realm of life, meaning, and purpose with the realm of space and time, cause and effect, mechanism and physical law.[4]

Shortly after Dennett's volume appeared in print, Steven Jay Gould wrote a negative review of it (see Gould, 1997a and 1997b), in which he summed up his opinion by saying something like, 'Dennett has produced yet another superficial book.' I now ask: Is the thought expressed in Dennett's claim (as Gould might have believed) something that is half-baked, superficial, and mistaken; or (as the many quotations of this passage apparently imply) is it an insightful statement of why Darwin deserves to occupy a central—and perhaps even the central—place in the history of thought? This question boils down to another, still simpler one: Did Darwin really succeed in unifying the two, previously unconnected realms of thinking and existence; or does it only seem to be true that he did this, to people who have not thought the matter through in a sufficiently clear way? There are many things I admire about Dennett. But readers will not be surprised to learn that I believe the second answer is correct.

Let me now change the subject slightly, by claiming that there existed at least one good precedent, in Darwin's own time and place, for another way he might have chosen to talk about his proposed metaphor of natural selection, as opposed to the way he actually did talk

---

3 What kind of thing am I talking about? The answer is that I have in mind experiments that are similar to the ones the psychologist Marc D. Hauser devised as a means of illuminating the thought and reasoning of animals. For example, see his book, 2000, *Wild Minds: What Animals Really Think*.

4 To mention a recent instance, William Calvin approvingly quoted this passage in his book 2002, p.14.

about it. One of Darwin's friends, the astronomer and philosopher John F.W. Herschel, famously referred, in a letter written February 20, 1836 to the geologist Charles Lyell, to ". . . that mystery of mysteries, the replacement of extinct species by others."[5] The suggestion I want to make here is that it would have been more satisfactory and honest—as well as something that later theorists would have found less confusing and troublesome—if Darwin had described natural selection in approximately the same terms as Herschel used to talk about the formation of new species.[6] Presumably, what Herschel meant when he spoke in the way he did was that it was difficult for humans like us to gain a correct understanding of the exact manner in which new species are created, and furthermore that, at the time of writing the letter, science still was very far away from being able to help us obtain an adequate grasp of that phenomenon.

Presumably, one of Herschel's motivations for expressing this opinion was that no person alive during his and Darwin's lifetimes could point to even a single unequivocally documented case where changes in animals, plants, funguses, etc., brought about by natural selection, subsequently led to the creation of a new species. (Neither, by the way—in spite of certain things Quammen says in his article—can any person do the same thing today.) To state the point more precisely, no one actually has observed a situation in which natural selection succeeded in changing some relatively isolated group of organisms in a way that resulted in the group in question no longer being able to interbreed with the larger group from which it had arisen, as opposed to its merely being transformed into a variety or subspecies of the larger group, and therefore still having the power to breed with other varieties of that same founding species. To mention a simple example, we are familiar with the idea that tamed wolves can be turned into dogs (which count as varieties and/or subspecies of wolves). But we have not yet observed any instances where natural selection has caused some particular breed of dogs to become so different from the wolves from which it descended, that it has been transformed into a separate species that no longer can interbreed with wolves at all.

---

[5] The philosopher Michael Ruse used this phrase as part of the title of one of his books. See his 1999.

[6] Biologists' technical term for the phenomenon of a species coming into existence is "speciation."

What is it about the notion of speciation that makes it difficult for humans to come mentally to terms with it? The answer, I suggest, is that we are only able to grasp and understand those things in a thorough and realistic fashion, which we repeatedly see, hear, feel, and otherwise deal with in everyday contexts. Therefore, whenever circumstances force us to deal with items that lie completely outside our experience, we tend to fall back on the crutch of reinterpreting and thus also transforming the unknown items from that which they literally are, into something else that we find more familiar, because of the fact that we have had more frequent experiences of the "something else." In my opinion, for example, anyone who represents natural selection (as Darwin sometimes did) as if it were a result, product, or upshot of genuine choices, has fallen into the trap of engaging in illicit, merely anthropomorphic reasoning, because it is not correct to say that any non-human creatures, or the whole of nature considered in a general fashion, is literally able to choose things in the sense of the word "choose" that applies to humans. Accordingly, it would be more forthright and less pretentious for us simply to admit that (similar to the case of speciation) scientific investigations have not yet provided us with any clear understanding of what natural selection literally is, or of the behaviors in which it is necessary for creatures to engage in order for them to realize and express that phenomenon, principle, or mechanism. In other words, in spite of our relatively high level of scientific sophistication, in which we take such pride at the present moment, it would be better for us to admit that both natural selection and speciation continue to be mysterious to us. In still other terms, it would be preferable in our present circumstances, for us not to give any answers to questions posed to us about those subjects, than to offer answers that are unjustified and misleading.

Over the long run, of course, scientists cannot be satisfied with mysteries. Nevertheless, it does not seem to me that people like E.O. Wilson are right to claim that, in principle, there are no mysteries left for present-day scientists to deal with, since they (and through them, potentially, all other human beings) already are in a position to give scientific answers to all factual questions without exception. (It seems to me the Plato-like idea just mentioned was the main thesis for which Wilson proposed to argue in his book *Consilience*, 1998.)[7] Furthermore,

[7] To be fair, Wilson draws a sharp distinction between that which he believes "in principle," and what he can claim to know in his role as a practicing scientist. For

we have no good reasons for supposing that theorists like Wilson are correct to suppose that the single, universally applicable scientific method we possess is the Cartesian one of breaking down each puzzling item into fundamental parts, and then discovering how to build up the original phenomenon again from those same parts.

Admittedly, it sometimes is correct and effective for us to use the method just mentioned as a means of solving scientific problems. For instance, I once heard someone say on a television science program that no present-day scientist is capable of accounting for the observed fact that our fingernails grow ten times faster than our toenails. Nevertheless, we can reasonably expect that future investigators sooner or later will discover a solution to this problem. Furthermore, it also would not be surprising if the method people used to explain this puzzle was to isolate and describe certain basic units out of which those bodily parts are physically constituted, and then to observe how growth processes affected those parts in different ways in the two cases, so as to make fingernails grow faster than toenails. But even having admitted all that, it still strikes me as an unjustified mistake to suppose that all scientific problems, and every potential solution to those problems, either are, or must be, similar to the ones just mentioned.

The following, favorite quotation of mine[8] describes a problem that seems to me to fall outside the scope and competence of what

---

example, he included a useful discussion of the so-called Biosphere project in the book just mentioned (pp.279-80), which seems to me to be a simple and elegant instance of the general point I am trying to make here about scientific mysteries. People assume that they are familiar with the workings of self-contained, life-supporting systems. Therefore they also believe that when humans arrive on some other planet, they will be able to create a system of that same sort under some kind of plastic bubble, and live inside it successfully and happily. To test that notion, such a system was built in the Arizona desert, and volunteers undertook to live there for two years, cut off from all outside influences except for electricity and communication. But it soon became clear to those who had organized the project that they had made some wrong assumptions. This fact was demonstrated by a long string of unexpected and bitter experiences—e.g. an explosive growth in the populations of cockroaches and ants, the extinction of all pollinating insects, the uncontrollable growth of weeds like morning glory vines, a falling off of oxygen and a great excess of CO_2, etc. In other words, it gradually became clear from all those disasters that they really had not been able to comprehend, after all, exactly how it was that nature "worked." (See Jane Poynter, *The Human Experiment: Two Years and Twenty Minutes inside Biosphere 2*, 2006.)

[8] What a repetitive bore I am! One indication of this is the fact that I also quoted this same passage in another place. See 2003, p.145.

Wilson seems to believe is a universally applicable scientific method. In this passage, physicist Richard Feynman reported how a puzzling but undeniable pattern of observations eventually forced people to revise the ways in which they previously had thought about very tiny natural objects and processes.

> "Quantum mechanics" is the description of the behavior of matter in all its details and, in particular, of the happenings on an atomic scale. Things on a very small scale behave like nothing that you have any direct experience about. They do not behave like waves, they do not behave like particles, they do not behave like clouds, or billiard balls, or weights on springs, or like anything that you have ever seen.
>
> [For instance,] Newton thought that light was made of particles, but then it was discovered, as we have seen here, that it behaves like a wave. Later, however (in the beginning of the twentieth century) it was found that light did indeed sometimes behave like a particle. Historically, the electron, for example, was thought to behave like a wave. So it really behaves like neither. Now we have given up. We say: "It is like *neither*." (1995, p.116)

Whether we are justified, in any given situation, in appealing to the Wilsonian scientific technique of analyzing, then reconstructing, phenomena in terms of basic units, is not a matter that should be settled by appealing to some a priori, metaphysical principle. Instead, it is a question whose answer—and the method of getting that answer—needs to be uncovered by means of empirical investigation.[9] *A fortiori*, it is also wrong to suppose that all scientific explanations must analyze their topics or objects (*"explananda"*) into units that most people find intuitively familiar. Although that may be true sometimes, and in some circumstances, it will not be true in others. Now let us ask: In what type of situations can one expect not to be able to explain things he or she finds puzzling, in terms of other things with which he is familiar? I suggest the answer is that this is likely to

---

[9] Isaac Newton assumed that his law or force of gravitation applied to all existing things everywhere and at all times. But Feynman reported that twentieth-century physicists subjected the supposed universality of this principle to empirical scrutiny, by means of astronomy. To be more precise, the individuals of which he spoke followed the procedure of looking further and further out into space, to determine—e.g., by noting the shapes of star clusters and of groups of galaxies—whether gravity still "worked" at ever larger scales, and thus in places that were extremely distant from us, and therefore possibly were different from our own situation. Feynman said that, at the time of his writing, Newton's idea so far had been confirmed, because the physicists in question had not yet found any areas where the principle of gravitation did not apply. (See 1995, pp.98-104.)

happen whenever someone tries to make sense of items that are totally new both to him and to every person who is similar to him.

I claim that the situation just mentioned, or something like it, not only applies to the behavior of very small bits of matter, but also to the biological phenomena of speciation and natural selection. Furthermore, one form of that same situation also applies to the problem we are considering in this book—namely, the question of how it is appropriate to think about human nature. In my view, the main reason we should not try to generalize to the human case, from facts about the natures of living creatures of other sorts with which we are familiar, is that humans—crucially unlike the case of all other known organisms—are "cultural animals." For example (to express the same point another way), we are not justified in supposing that we always will be able to make sense of humans' distinctive nature just by means of appealing to Darwinian natural selection, because attempting to do that would leave out of account most of the important influences that history, tradition, and culture have had—and to some extent, still are having—on what we now are like.

2.2 Why culture is real[10]

One problem with trying to answer the question of what makes human beings different from all other creatures is finding a way to deal with the implications this question has about the topic of existence. For example, consider the following paragraph from Terrence Deacon's book, 1997 (pp.21-2), which sketches a theoretical picture that is at least somewhat similar to the one set forth in the previous section, because it apparently agrees with the claim I made there that doing things for reasons (as humans usually but not always do) is not the same as doing things from other causes of behavior.

> Though we share the same earth with millions of kinds of living creatures, we also live in a world that no other species has access to. We inhabit a world full of abstractions, impossibilities, and paradoxes. We alone brood about what didn't happen, and spend a large amount of each day musing about the way things could have been if events had transpired differently. And we alone ponder what it will be like not to be. In what other species could individuals ever be troubled by the fact that they do not recall the way things were before they were born and will not know what will occur after they die? We make use of these stories to organize our lives. In a real sense,

[10] My student, Jason Kennedy, suggested this title.

we live our lives in this shared virtual world. And slowly, over the millennia, we have come to realize that no other species on earth seems able to follow us into this miraculous place.[11]

Even if Deacon's paragraph throws some light on what it means to speak of humans as cultural animals, it also introduces new problems of its own. In particular, his talk about a separate world open to humans, but not to other creatures, is obviously a metaphor. The fact that it is a metaphor temporarily excuses him from having to answer strange, embarrassing, and seemingly unanswerable questions like the following: (a) "Where, when, and how does the 'world full of abstractions, impossibilities, and paradoxes' exist?" (b) "What does it mean to suppose that impossibilities exist, when most people assume that part of what it means to say that some event, fact, or object **X**, is impossible, is to convey the information that **X** does not exist?" (c) "Since nearly all practitioners and interpreters of modern science accept the idea that there is just one world rather than several,[12] is anyone who accepts what Deacon says here also committed to rejecting the outlook and procedures of science?" As far as I can see, Deacon never faces the task of explaining these ontological implications of his view, since he does not provide any directions about how it is appropriate to interpret his metaphors literally.[13]

Even though Deacon does not talk about the connections human nature apparently has to existence, the same is not true of some other philosophers. For instance, some remarks Noam Chomsky made on this subject provide an instructive contrast with Deacon. In Chomsky's view, speculations about language, as well as other parts or functions of the human mind, need to be closely connected with ongoing scientific work. For example, he appeals to this point in order to justify the fact that his general conception of things (prominently including language) has undergone several radical changes over the

---

[11] I quoted this same passage in another book. See 2003, p.11n5.

[12] John Searle begins one of his books with the sentence: "We live in exactly one world, not two or three or seventeen." (1995, p.xi) My own view is similar to one aptly expressed by the philosopher Mary Midgley in the title of one of her papers (1998)—"One World, but a Big One." (I am indebted to Jason Kennedy for drawing my attention to the work of Midgley.)

[13] A similar problem is connected with another metaphor Deacon introduces in his book—namely, the idea that young children find it easy to learn languages because language is a "virus," which gradually has configured itself in such a way as to be able to infect minds of their type in an especially efficient manner. But this last problem is another story, with which I shall not concern myself in this book.

course of his career. One theme that has been fairly constant throughout all these changes is his claim that the mind is something physical. To illustrate this point, in a fairly recent version of his account (see 1997a, especially pp.17ff., and 1997b passim), Chomsky apparently accepts the truth of all four of the following claims. (i) Scientists' job is not to deal with mere abstractions, but to examine, describe, and explain existing things. (ii) Everything that exists occupies a single definite position in space and time. (iii) The study of language, as well as mind considered more generally, is—or at least should be understood as—an ordinary part of objective, empirical, and naturalistically unified science. Thus it follows that (iv) neither any human language, nor any of the parts, structures, or rules that underlie and constitute such a language, is something that is general and abstract. Instead, languages and all their parts are entirely composed of concrete physical objects that are particular rather than universal, and that have completely determined spatial and temporal locations.

What physical objects compose language? Where, when, and how, according to Chomsky, do those objects exist? His answer is that all the things just mentioned are biological structures and processes that are identical with various parts of the brains of language users. He says, for instance, that there is no such thing as the Italian language considered in an abstract sense. Instead, there is only the particular and concrete version of Italian that exists inside the brain of Marco, the particular version that exists inside the brain of Maria, and so on. Therefore, what it means to speak of Marco, Maria, and many other people as being speakers of the same one language—Italian—is simply that those people's brains happen to be structured in sufficiently similar ways, to enable all of them to communicate verbally with one another over some non-trivially large range of discourse.

One basic source of the position just summarized, I suggest, is Chomsky's skepticism about the explanatory value of the so-called social sciences, as opposed to what he refers to as the "core," natural sciences. Social scientists like anthropologists, economists, sociologists, archeologists, and psychologists often claim to have discovered facts, laws, etc. about the world. But in Chomsky's view, we are not justified in accepting any of those purported discoveries as a legitimate part of the body of scientific knowledge, unless and until someone shows that one can reanalyze and re-express that discovery in such a way as to make it consistent with—in fact, a mere extension of—the laws and principles that belong to natural sciences like physics and chemistry. So far, according to Chomsky, there is only one social science whose practitioners have made a significant amount of progress in the program of integrating their field with the natural sciences. That social science is linguistics—the one in which he himself is an expert. Strictly speaking, he says, linguistics counts as part of the more encompassing science of psychology, whose job it is to study humans and their behavior. Nevertheless, linguistics has a special status and importance within psychology, by virtue of the fact that the phenomena with which it is concerned—language, and humans' language-using behavior—have a unique level of completeness, detail, and complexity that far surpasses the complexity of the phenomena that are studied by any other of the social sciences.

Although Chomsky is an intelligent, honest, and thoughtful person, and also in spite of the great amount of fame and influence he enjoys at the present moment, my impression of the doctrine just described is that it is such a strange, unrealistic, and implausible one as to be silly. Let me begin to explain what I mean when I say this, by mentioning something I take to be a piece of evidence that supports it. I was excited when I first read in the writings of Ian Tattersall and other paleoanthropologists about a particular historical fact their research had uncovered. To be specific, their investigations showed that (1) our relatively early, direct ancestors shared many cultural traditions (e.g. particular ways of manufacturing and using stone tools) with members of other species of hominids with whom they lived at the same time, and to whom they were related. Furthermore, (2) because of the thorough intimacy of that sharing, the developmental path cultural evolution followed during those times was almost completely independent of the coming and going of different species of hominids. For example, for more than half the time our species, *Homo sapiens*, has been in existence, some of our conspecific ancestors lived in close proximity (in the Levant, the eastern shore of the Mediterranean) with members of our nearest, "cousin" species, *Homo neanderthalensis*. Moreover, our sapiens ancestors acted in ways that were nearly indistinguishable from the behavior of their Neanderthal neighbors. Thus, Alan Thorne and Milford Wolpoff (see their 1992, pp.77-8) report that our ancestors not only made the same types of stone tools as their Neanderthal contemporaries did, using the same manufacturing technologies, and doing so at the same frequencies, but both of those groups also shared the same stylized burial customs, hunted the same game, and even used the same butchering techniques in the preparation of that game. In addition, those two groups also made and wore clothes of the same type; they both made fire using the same techniques; and they lived (at different times) in many of the same sheltering caves. (For context, references, and further details about this topic, see my 2003, pp.14ff.)

Nevertheless, all these historical facts seem to be incompatible with the metaphysical-like view that Chomsky accepts. This is true because, whenever a new hominid species is created, then—as part of that same event—a new sort of hominid brain also comes into existence. And if Chomsky were right to suppose that thinking always is exhaustively determined by the detailed physical structure that belongs to the brain of the person who does that thinking, then that would imply that the owners of the brains of each of the new species of hominids immediately should have started to act and think in new and different ways. Yet—as just mentioned—the evidence does not support this idea. Still more generally, nothing in the fossil record indicates that there was any sort of "jump" or "zig" in behavior, each time some novel hominid speciation took place. Instead, observations show that what repeatedly happened is that the members of each new species of hominids continued to think and act—at least to begin with—in the very same ways as their immediate ancestors (of different species) had done before them.

This empirical discovery about the general pattern of the lives of our early relatives is a clear instance of scientists going about their proper business of informing us about what the world is (and was) like. But would this discovery have been clarified, would it have been made more securely grounded and justified, and (*a fortiori*) would it have been transformed from pseudo-science into genuine science, if—following Chomsky's pronouncements—someone had succeeded in reformulating it in such a way as to make it take account of particular brain structures, or particular patterns of neurons, or even of individual groups or sets of molecules that existed inside the heads of our ancient forebears? As far as I can see, the answer is no. More generally, I think the basic problem with Chomsky's program of supposed "genuine explanations" is that discoveries like the one just described are not aimed at, nor are they even relevant to, the topics of brains, neurons, and molecules. Instead, those discoveries are only concerned, simply and literally, with the social and cultural conditions that once existed in various ancient populations of primates of the special, two-legged sort we now call hominids. Therefore (contrary to Chomsky's dictates), to reinterpret and re-express the previously mentioned discovery in terms of physical objects like brains and neurons would not strengthen, legitimize, or in any way "fix" it. Instead, it merely would change the subject.

The deeper point that lies behind what I just have been saying is that it always is wrong to suppose that appeals to the notion of existence will (or can) make valuable contributions to the worth of scientific explanations—especially if the sort of existence in question is only a product of some programmatic definition, as opposed to being a result of actual observations. For example, what leads Chomsky to say that researchers could not legitimately claim to have discovered anything of scientific importance about human language, and about how language works, unless and until someone reformulates that discovery in terms of the contents of the brains that belong to language users? I do not think he was forced to accept the conclusion that certain brain contents always determine the forms and contents of language, because of something he learned on the basis of observations. Instead, it seems to me that he has conceptual rather than empirical reasons for saying this. But if that is so, then it follows in turn (let me apologize, if what I am about to say sounds like a flippant denunciation of someone I admire) that we are dealing here with a piece of ideology or a priori metaphysics, rather than a part of genuine, empirical science.

2.3 It is misleading to suppose that the existence of cultural items depends on conscious stipulations

> Lack of historical sense is the family failing of all philosophers. Only that which has no history is definable.
>
> Friedrich Nietzsche

We are not the only earthly creatures that have and employ cultures. Furthermore, some of the cultural practices in which non-human animals engage are roughly analogous to corresponding ones of ours. This point might seem to imply that that our species differs from other species only in degree but not in kind. However, although some people (e.g. Barbara J. King) choose to speak in this way, it still remains a fact—as Wilson once observed (see his 1975, p.272)—that the gulf between humans like us and even those animals that apparently are most similar to us is such a large, complex, and many-sided one, that we know of no scale capable of measuring it. In view of this last point, it is arguably true to say that in spite of the fact that humans are not separated from other animals by being the sole inventers and users of culture, our species still counts as special by virtue of (so to speak) the amount of culture we use, and the extent to which we use it. To be more specific, our species' distinctiveness stems from the fact that, unlike other creatures, culture permeates virtually everything we do, and therefore has a far more powerful influence in determining what we are like—our nature—than is true in the case of animals of any other known type (see ibid., p.284).

Let me now begin to sketch a more detailed picture of what present-day scientists have learned about non-human cultures analogous to ours, by mentioning a case I heard of a long time ago, involving rhesus monkeys (see Ardrey, pp.45-6, 88-90, 107-10). A group of experimenters under the direction of C.R. Carpenter was given access to a small island of thirty-six acres close to Puerto Rico in the Caribbean, as a place to conduct research on animal behavior. Their experimental procedure was to release 350 unrelated rhesus monkeys drawn randomly from different areas in India, onto the island at the same time, and wait to see how those animals would organize themselves. What they discovered was that, over the course of about a year, the monkeys formed several troupes or bands, and divided the island into exclusive territories, each occupied by one of those bands. Moreover, each band constantly and vigorously defended the borders of its territory against incursions from neighboring bands. Still later, these scientists also began to notice certain group behavioral patterns that were connected with individual monkeys. For instance, in one of the bands there happened to be a male that was especially large and aggressive. Whenever experimenters removed that monkey from its troupe, they noticed that, soon afterward, the borders of the troupe's territory receded and became more constricted. But when they returned the monkey to its former place, the borders expanded again. I find this observation interesting, because (i) leaders and role-models cannot change any of the bodily characteristics of their followers, and (ii) it would be physically impossible for any single monkey to participate in the defense of all the borders of its territory, at the same time. Presumably, therefore, (iii) the presence of this particular monkey in the troupe had a "galvanizing" effect on all, or at least many, of its fellow troupe members, which made those other monkeys more confident, brave, and combative than they otherwise would have been. Furthermore, (iv) this particular case seems to have obvious parallels to the careers of inspiring human leaders like King Cyrus of Persia, Alexander the Great, Julius Caesar, Jesus, Mohammed, Genghis Kahn, Martin

Luther, Napoleon, and Lincoln. In fact, I think of it as one more, fairly standard illustration of the truth of the familiar Napoleonic dictum that "in war, the ratio of the moral to the material is 3 to 1."

What conditions allow cultural traditions to spread from one human being to another, and from one human generation to another? At the present time, the ideas most people have about exactly what those conditions are, and how they work, are rough, approximate, and simple-minded. At one stage of my life, I received a graphic confirmation of that point. During the academic year of 1971-2, my wife and I moved our young family from Toronto, Canada to Oxford, England, for a first sabbatical leave from York University. Our two daughters, Kirsten and Sigrid, were then 2 and 4 years old; and by the time we returned to Canada at the end of our stay, they were 3 and 5, and we also had produced a third daughter born in England who, again by the time we left, was not yet 1. Because our older daughters were offspring of a pair of American landed immigrants based in Canada, and since we knew they would be spending far more of their time with us than with anyone else during their stay in England, I assumed in a thoughtlessly naïve fashion that they would continue to act and speak in much the same manner as my wife and I did. (The third daughter, Erica, born in November of that year, was not relevant to these thoughts since, during the whole time, she still was too young to speak.) Nevertheless, it soon became clear that this expectation was false, foolish, and unfounded, because the humans with whom I was dealing were "programmed" in a different way than I supposed. In particular, even though our eldest daughter (and, to a lesser degree, our second daughter) came in contact with only a relatively few models of what I considered to be a pleasantly refined, but slightly pedantic, academic Oxford accent,[14] they appeared to absorb that way of talking almost overnight. Moreover, along with the accent, they also acquired, just as rapidly, a whole new suite of mannerisms, attitudes, and what seemed to be strongly held personal beliefs and values. For example, our oldest daughter Kirsten (the same person who later produced the cover picture for this book) soon became embarrassed by the way she heard her parents speaking, and occasionally presented little lectures about what she thought we might do to solve that problem. Let me try to convey some of the style as well as content of one of those lectures, by quoting as follows.

---

[14] It is appropriate to add, however, that inhabitants of Oxford frequently expressed the opinion that it was we, not they, who had an "accent."

"Oh Dad-dy! Don't say 'pants'! 'Pants' is the *ug-gly* wuhd. 'Trouwszas' is the *beau-ti-ful* wuhd!"

Eventually, however, all this came to an end. In no more than two weeks after our arrival back in Toronto after a year's stay, the Oxford accent disappeared, because our children then had switched to copying the Central Canadian model instead (still a somewhat different item from the Western and Mid-western American accent my wife and I habitually spoke). Nevertheless, occasionally there were "flashbacks." The last of those I remember happened in late November or early December of that year, when we took our children on a night trip through a part of the city that had especially impressive, lighted Christmas decorations. At one point, Kirsten exclaimed enthusiastically that the colored lights we were seeing were "smashing"!

Consider a related question. Is being exposed to differently acculturated people enough, just by itself, to bring about some type and degree of cultural transmission? Even if there is a sense in which the answer to this question is yes, it also is clear that mere exposure is not a sufficient condition for acquiring cultural traits, because people are able to choose, or in one way or another they can be forced, to consider certain influences as "desirable and proper" and others as "improper and undesirable." As mentioned before, for instance, even though our children in England were in the presence of my wife and me and other visitors from North America for a much longer time than they were exposed to their local English friends and acquaintances, they did not aspire to talk like us. Furthermore, still another sign that the processes by which one assimilates cultural influences are not entirely passive and impersonal is that factors like age, intelligence, attitudes, and talent play a role in them as well. For example, although some people (especially children under 4 years old) pick up a new language rapidly and without a great deal of effort,[15] other individuals may be exposed to the sounds of a foreign language for a very long time, but never manage to "get the hang of it." Thus, some obviously intelligent people, like Henry Kissinger and Arnold Schwarzenegger, who have

---

[15] Age is not the only thing that operates in such cases. For instance, I once heard Chomsky say in a recorded lecture (in a series entitled "Language and Mind") that he personally was acquainted with at least one adult who could recreate a child-like mental condition in himself, that allowed him—as presumably he also had been able to do when he was very young—to obtain a working grasp of a new language in as short a period as two weeks.

lived most of their lives in an English speaking environment, continue to speak English throughout their lives with a strong foreign accent from the language (in these two cases, German) they learned as children. It is not clear to me how to explain this observed fact. For example, I do not know whether it is correct, or even meaningful, to try to account for such a phenomenon by saying that people of that sort "lack an ear" for language.

One paranoid group who feared that simply being in the presence of individuals of a particular type was enough to suffer cultural contamination from them was certain leading individuals involved in the Bolshevik Revolution in Russia, during the First World War. The members of that group wanted Vladimir Lenin to return to Russia from exile in Switzerland, so he could assume a prominent role in directing the political unrest that had broken out in his homeland. But in order for him to come back, he would have to travel across the territory of the hated German enemy whom the Russians just then were engaged in fighting. As a solution to this problem, they negotiated with various German authorities (who also were eager for Lenin to return, since they believed—correctly, as it turned out—that he would take the Russians out of the war) to allow Lenin to make this trip in a sealed boxcar. By this means, both the Bolsheviks and the Germans reasoned, he would be able to stand before the Russian masses and truthfully declare that he had nothing at all to do with Germans, in spite of the fact that he had been forced to return to Russia by traveling across their country.

Luckily for those members of the Bolshevik party, all of them were born too early to hear a paper delivered at one of the sessions I attended of an annual meeting of the *Southern Society for Philosophy and Psychology,* whose topic was a series of meditation experiments conducted by psychologists from the Maharishi University in the United States, in the American city of Detroit. As the authors of the paper explained, those experiments took the form of psychologists traveling to Detroit, which then was the city with the highest crime rate in the country, and furthermore going to the particular part of Detroit that currently had a higher crime rate than any other, and engaging in group meditations on the themes of peace, love, justice, forgiveness, etc. for several days, in a sealed room with covered windows, and then checking later to see whether the crime rates for that area had dropped during the time of the meditations, in a way that was sufficiently substantial not to be attributable to chance. The main claim the authors made in their address was that, in fact, the rates had dropped in exactly the way I just have described.

I was disturbed and unsettled at the time when I first heard about those experiments. As I listened to what the speakers were saying, I kept asking myself how the organizers of the conference could have been so irresponsible as to have accepted such a mad, fraudulent, and pointless paper, for presentation to an intelligent and relatively informed audience like myself and the people sitting around me. After all, there was no possible way in which the meditations that took place inside the closed room could have affected the thoughts, motives, and emotions of the inhabitants of Detroit who happened to be relatively close by, but outside the room, since those inhabitants did not know—and would not have cared even if they had known—that those meditations were going on! Later however, after having calmed down, I began to think the matter through more clearly. I then concluded that something like the following scenario must have taken place. The conference organizers (or people they trusted) must have checked the Detroit crime statistics for the time in question. And since checking the actual numbers showed that the data in fact did correspond to, and therefore did support, the precise and limited claim the psychologists were making in their paper, it would have been arbitrary and unfair for the organizers not to accept this paper for presentation, even if none of them was able to think of any reason that the data should have been what they were. At any rate, the only point I want to make at this point in the chapter is that considerations of the sort about which we now have been speaking at least should force future historians of the early twentieth century to think about the case of Lenin's sealed boxcar in a slightly different fashion.

Let us now move on to another matter. I propose to argue in the remainder of this book that ancient cultural traditions have played an important role in determining what human beings now have become. However, not many other present-day philosophers of mind, psychology, and biology have shown a great deal of interest in the general topic of culture. One philosopher who does have such an interest is John Searle of the University of California, Berkeley. In fact, it might be correct to say that his special way of thinking about things like laws, attitudes, marriages, songs, commands, bids at an auction, aspirations, and insults has become the most popular philosophical view of culture and cultural items at this present moment. Nevertheless, in my opinion, even though Searle correctly has assessed the importance of this formerly neglected study, there are some respects in which the kind of account he proposes is unrealistic and misleading, and needs to be replaced by a theory of a different sort.

The principal example of a cultural item that John Searle discusses in his book, *The Construction of Social Reality* (1995), is money. He presumably made that choice because he considers money to be a simple, non-misleading, and naturally extendible case of the topic of the book. In particular, Searle emphasizes the point that cultural items come into existence through accepted, agreed upon, but essentially arbitrary, human choices; and therefore it is appropriate to believe that they are "thought items" rather than physical objects. According to him, money provides a particularly good illustration of this idea, because—as world travelers know—the pieces of plastic and paper we have in our wallets and checkbooks, and the flat, circular pieces of metal we have in our pockets, cannot constitute money just in and by themselves, as proved by the fact that, depending on place, time, and circumstances, people sometimes will and sometimes will not accept them in payment for goods and services.

It seems to me that Searle's view of money is analogous to the "social contract" theory of government once proposed by the French-Swiss philosopher, Jean-Jacques Rousseau. Rousseau claimed that the way states come into existence was by people (a) calculating their self-interest, (b) coming to agree among themselves that they would be better off as citizens of a state than in the simpler, state-less societies in which they currently lived, and then (c) making a rational and voluntary decision to do away with those simpler societies. However, as Jared Diamond points out (1999, p.283), observation and historical records have failed to uncover even a single case of a state's being formed in that ethereal atmosphere of dispassionate farsightedness. More generally speaking, Diamond says it never is the case that smaller societies voluntary abandon their sovereignty and merge into larger ones. Instead, this only happens as a result of conquest or of some other form of external duress.

I agree with Diamond that people do not create governments by means of voluntary choices, or by reasoning about their mutual advantage; and I want to argue that something similar also applies to John Searle's example of how we create money. One point that shows the parallelism of these two cases—at least in an indirect fashion— is the role precious metals like gold, silver, and platinum have played, and continue to play, in the history of money in Western and Eurasian Society. Of course, if and when people discover some cheap way of producing those metals in a laboratory or factory, their function of being bearers and standards of monetary wealth will be finished. But until that happens, it is instructive for us to reflect on the question of what has made them important for such purposes in the past, and what preserves that importance at the present time, in spite of efforts the leaders of some modern governments have made to disparage and undercut that influence.

Let us take gold as our particular example. I claim that the monetary importance of this metal is not just based (as Searle evidently supposes) on some authoritative person's having made a conscious "speech act" of stipulating that gold will perform that sort of monetary function. Rather, it also depends on a certain kind of historical memory that was formed, more or less unconsciously, by a very large group of humans. There is a saying one sometimes hears at the stock market: "During times when all myths die, only the oldest of them survives." What is the "oldest myth" to which this saying refers? It is the myth of eternal life, which is represented and expressed by this particular chemical element, which never appears to change its properties. For example, in a very old tomb or grave, one can expect to find one or more bodies that are so decayed that almost nothing is left of them. (The teeth usually go last.) In contrast with this, however, a lump of gold that once filled a now pulverized tooth, or a golden ring that loosely encircles some dusty finger bone, will look as if it had been dug up and refined yesterday. Thus, an important part of the reason Eurasian people came to value gold, especially as part of grave goods, is that they associated it (mainly in an unconscious fashion) with eternity. To be even more specific, gold's power over us largely stems from the fact that—for a very long time—people in our society (but admittedly, not people in some other societies) have accepted gold as a symbol of the ancient, human religious hope that something connected with themselves also might succeed in outlasting all of the changes of time.

Accordingly, there is something wrong with the Searlean claim that cultural entities—once they have attained the status of being powerful influences in our lives—cannot have anything more than the kind of existence that belongs to a word that some person arbitrarily and perhaps even accidentally proposes as a name for some new element, process, or organism that he or she has discovered. Instead, things of this general sort can and very often do also have a certain kind of objective, physical, and non-legislated existence, in light of the fact that many of the cultural factors that influence us most strongly are not items of which any individual ever has managed to become conscious. In this respect, I agree with Chomsky who maintains, in opposition to Searle, that in the case of both cultural and non-cultural things alike, definability is only a mark of "play" or "pretended" existence as opposed to genuine existence. Also, it seems to me that the quotation from Nietzsche with which this section began deserves to be taken seriously: "Only that can be defined which has no history." Both of these ideas serve as useful antidotes to the notion that cultural items only can exist if, when, because, and for as long as, someone knows, says or believes that they exist.

I want to end this chapter by pointing out a connection between my criticism of John Searle's proposed account of cultural and social reality and the following sentence from an old Mother Goose rhyme: "If wishes were horses, then beggars would ride." Searle claims that, since money is not the same thing as the objects—e.g. pieces of paper and metal—that happen to represent and embody it, it therefore follows that money cannot have any more of a genuine existence in the physical world than a wished-for horse. But it also is possible to think about that same rhyme in slightly different ways. And each of the ways I now am going to mention shows how a cultural item might exist and might be capable of acting in physical terms, in spite of the fact that it is not something (like a horse) on which one is able to put his hands, and whose being is localized at definite points in space and time, and which thus exists entirely outside of our skins.

Here is my list of new ways in which one can think about something like the Mother Goose rhyme quoted before: (i) "If wishes were planning, then beggars would make plans." (ii) "If wishes were decisions, then beggars would decide." (iii) "If wishes were serious resolutions to seek redress for past wrongs, then beggars would start to obtain justice." (iv) "If wishes were concrete steps towards behaving in an open, sympathetic, and approachable manner, then beggars would make friends." Of course, I do not deny that there is an obvious difference between items that we can touch with our hands, arms, and legs, and various changes in the brains, nerves, and muscles of one or several people that it would be difficult for us to think of as "tangible." Because of that difference, I admit that it never has happened, nor will it ever happen in the future, that a beggar has been able to create, simply out of one or more of his or her wishes, a horse that he or she is able to saddle, mount, and ride. Nevertheless, some people of the general type we now are discussing, on some occasions in their lives, have done and will do things like making plans, making decisions, changing their lives in such a way as to seek redress, and taking steps[16] towards making friends. Because of this fact, it is plausible to claim that any person who succeeds in doing something like the things just mentioned, partially has succeeded in transforming himself into something more than just an impotently dreaming, constantly frustrated beggar.[17]

One reason I subscribe to the view that not all cultural items are things of which people are, or even could be, completely conscious is that those items can—and often do—reach far back into the past, and extend far into the future. Still more particularly, it is not appropriate to appeal to introspection as a means of answering the question of exactly what some cultural entity is and exactly how it operates. Instead, one should attempt to settle questions of that sort by taking account of such indirectly but nevertheless objectively observable items as (e.g.) a person's behavior, his or her personality, and his past history of triumphs and failures, as reflections and indications of his personal strengths and weaknesses.

---

[16] This points to a possible answer to Gilbert Ryle's famous philosophical argument about "willing to will." Ryle said (see 1949, Chapter III, especially, p.67) that if certain philosophers maintain that every voluntary action is preceded by a volition, and if they also assume that performing a volition is something a person does voluntarily, then it would be necessary to perform volition number minus 2 in order to perform volition number minus 1, perform volition number minus 3 in order to perform volition number minus 2, perform volition minus 4 to perform volition minus 3, and so on *ad infinitum*, and therefore one never could perform any voluntary action at all. My suggested solution to this puzzle is that one can identify the so-called volition that leads to an action done on purpose, with the first steps, or at least the first part of the steps, that a person takes to perform that action voluntarily; and because of that, doing the action would not have to involve a vicious regress after all. (I owe this general point to Wilfrid Sellars.)

[17] Another dictum one hears at the stock market is Gresham's Law that "bad money drives out good"—i.e. the idea that people hoard the particular kind of money they trust, and spend (try to get rid of) money they do not trust. It strikes me as obvious that trust is crucial to the success and legitimacy of things like money, marriages, governments, and dance partnerships. But real (i.e., "good" and "trusted") cultural items are different in kind from unreal ones, because the real ones have connections to and correspondences with actual, observed, historical factors, in a way that unreal ones do not. For example, people's trust in a cultural entity like a piece of money or a particular government is something that is physically constituted by sets of habitual connections that those people have formed, on the basis of experience and learning, between the money or the government in question, and various items in the physical world.

Chapter 3

One Invention that pointed the way toward Present-day Human Nature: The First Domestication of Animals

> Animals are not brethren, they are not underlings. They are other nations, caught with ourselves in the net of life and time.
>
> Henry Beston

> Civilization is based, not only on men, but on plants and animals.
>
> J.B.S. Haldane

3.1 Instead of beginning a review of our species' most important properties by talking about the complex and mysterious ability to speak, it is clarifying to focus first on the simpler, and earlier acquired, ability to tame and exploit some of our fellow creatures

What would you guess is the most numerous and common type of bird that now inhabits our planet? Is it the sparrow—e.g. the House Sparrow? Or the starling? Or the seagull—e.g. the Herring Gull? If a small note that appeared in my local newspaper is to be believed, the correct answer (by a very wide margin) is the chicken. The anonymous author of this short article from whom I learned this *prima facie* fact did not have in mind the Prairie Chicken, of various species, which is listed in the index of my copy of Roger Tory Peterson's *A Field Guide to the Birds* (1964, p.277). Instead, the subject of the article was the domesticated chicken, comprised of familiar varieties like Rhode Island Red, White Leghorn, etc., whose meat one finds for sale in grocery stores. In fact—again according to this same source—humans eat the incredible number of about 30 billion of these birds each year, more than 1,000 of them per second. But in spite of points like these, Peterson does not see fit to discuss them in his book.

There are at least two considerations that might lead professional ornithologists like Roger Peterson not to pay attention to domesticated chickens. The first is that such authors might be influenced by the fact that chickens are closely associated with human beings. This fact might lead them to reason that since it is not appropriate to think of humans as parts of the natural world, it must follow that domesticated animals like chickens also cannot belong to that same world. However—in view of the points that (i) there only is one planet Earth, and (ii) at the present time humans and their activities take up a very large part of the energy and space of that planet—does it really make sense to suppose it is possible to draw a line between the natural world on one side, as opposed to something else on the other? Another anonymous article from my paper describes a current dispute about this question, in which British geoscientists now are engaged. Some of those scientists say that the changes wrought since the Industrial Revolution 200 years ago are so profound that they have left visible marks on the physical and living fabric of our planet. Therefore, according to them, it would be appropriate for us to recognize a new epoch in the official geological time scale—one based on human activity—known as the "Anthropocene" era. As opposed to this, others argue that such an idea is nothing more than an expression of the vanity of our species; and there is no justification for creating a geological epoch that describes some single historical event, however long- or short-lasting that event might happen to be.

A second point that might have led ornithologists not to take account of the existence of chickens is that those birds and their eggs only are brought into being in the first place in order for them to be killed and eaten very soon afterwards. Thus, chickens are apparently very little else than living products that humans have created (using various unnatural and artificial means to do so), in order to serve selfish purposes of their own. For instance, it does not seem to be true that chickens, considered in themselves, have any independently sufficient reasons for existing, or any future state towards which natural processes are causing them to evolve.

Again however, several points throw doubt on the usefulness of this last way of thinking about domesticated creatures. For one thing, the great majority of wild birds also apparently live in much the same fashion, and on something like the same scale of time, as chickens do. For instance, I remember reading in what struck me at the time as a credible source (but I apologize for not remembering what source it was) that, on average, of all the many billions of wild birds alive on the Earth at any particular moment, fully half of them will be dead just one year later.

Still another factor that might mitigate the force of such a dismissive attitude towards chickens is that humans, employing their so-called artificial methods of domestication and selective breeding, are not the only creatures who have entered into symbiotic relations with organisms of other species. For instance, the biologist David Attenborough points out that certain species of ants raid the nests of ants of other species, and carry off the pupae they find there. When those pupae hatch, the young ants then serve their captors, by collecting food and feeding it to them. Attenborough points out that at least some of the "slave-making" ants are forced to operate in this fashion, because those particular ants have jaws that are so large that they cannot feed themselves. (See Attenborough, 1979, p.104.) In another book written by the same author (1984, p.116) he describes ants of still other sorts that have discovered a way of getting access to the nutrients contained in grass, by using domesticated aphids as intermediaries. Aphids are very tiny insects that only digest a small part of the sap they regularly suck from grass leaves, and excrete the rest as a sugary liquid known as honeydew, which ants of that species eat and digest in turn. Thus, the ants' relations with the aphids turn out to be similar to those of human dairy farmers with their cows. To be more precise, the ants collect the aphids into large herds, which they protect from other insects that invade their grazing area, by squirting the invaders with formic acid. Then they follow a regular schedule of "milking" the aphids, in order to nourish themselves with the resulting honeydew. Attenborough adds that some of these farmer-ants even encourage their aphids to produce more honeydew than they normally would do, by stroking them repeatedly with their antennae.[1]

In the previous two chapters, I already suggested various respects in which it might be correct to think of human beings as different from all other animals. For example, one expression of the separation of our species from creatures like dinosaurs, crocodiles, snakes, whales, koala bears, fish, etc. is the fact that archeologists are not able to reconstruct, and thereby know, the style of life that once belonged to any particular human whose bodily parts they excavate, just by examining those remains themselves.[2] Accordingly, anthropologists

---

[1] How long have ants been engaging in such behavior? Another anonymous short article from my newspaper ("The First Farmers" *The Globe and Mail,* Monday, April 07, 2008, p.L6) informs us that some ants were practicing a highly sophisticated form of farming, at least 50 million years before the Sumerians and Ur people of the Fertile Crescent discovered agriculture.

[2] Consider the following passage from the last part (p.115) of Ernest Hemingway's novel, *The Old Man and the Sea*:

need to examine more than just the remains of human bodies in order to figure out how those humans once made their living. They also must make inferences from any physical artifacts they happen to find in the same general area, on the assumption that those artifacts were "extensions" of the bodies of the humans who made them.[3] Furthermore, archeologists need to look for evidence of domesticated plants, animals, and fungi in the vicinity of the human skeletons they excavate, because those domesticates are "once living artifacts" that the past humans created by gathering or capturing, and then selectively breeding, their wild progenitors.

Over the last few centuries, experts have amassed a large group of facts about changes that take place, at fairly predictable rates over time, in domesticated plants, animals, and fungi, as compared with their wild ancestors. Let me mention three simple examples of facts of that sort. First, for still unknown reasons, the cores of the horns of semi-domesticated goats tend to be almond shaped in cross section, while cross sections of the horns of goats that have descended from many generations of domesticates are flat on one side; furthermore, the horns of goats that are products of even longer periods of domestication have cross sections that are kidney-shaped. Second, the bones of all domesticated animals, as compared with those of their wild relatives, tend to become progressively less dense as those animals pass through succeeding generations. Finally, the sizes of domesticated animals also tend to change in predictable ways. As a general rule, whenever animals that are smaller than a chicken or a rabbit

---

He could not talk to the fish anymore because the fish had been ruined too badly. Then something came into his head. "Half-fish," he said. "Fish that you were. I am sorry that I went too far out. I ruined us both. But we have killed many sharks, you and I, and ruined many others. How many did you ever kill, old fish? You do not have that spear on your head for nothing." He liked to think of the fish and what he could do to a shark if he were swimming free.

[3] Paleoanthropologists sometimes dream of finding the undisturbed remains of a whole tribe of Upper Paleolithic or Neolithic humans, whose members were making and employing artifacts of many different sorts, until the time when all of them suddenly were smothered by a massive mudslide. (They imagine this hypothetical event as analogous to the occasion when ash and lava spewing from the volcano Vesuvius quickly covered—and thereby preserved—the Roman towns of Pompey and Herculaneum along with all their inhabitants in the year A.D.79). But so far at least, that dream has not proved to be anything more than an idle wish.

become domesticated, they become larger as compared with their wild relations, while those that originally were larger than a chicken become smaller.[4] Thus, by paying attention to matters such as these, archeologists examining the remains of animals of other species that were found fairly close to the bones of long dead humans, can determine whether those animals were domesticated, and if they were, for approximately how long their ancestors had been domesticated before them.[5]

It might seem easier to understand how past humans learned to begin planting and harvesting useful food plants like emmer wheat, or edible fungi like common (*Agaricus campestris*), shitaki, or portobello mushrooms, than to see how they first got the idea of domesticating wild animals. Imagine, for example, that a group of early humans living in the Fertile Crescent—e.g. in a place that is now part of present-day Iraq—used bone sickles with flint edges as tools for gathering a supply of grain from a ripe stand of wild wheat, which they then stored in a pottery vessel. But on the way home, one of them accidentally spilled the vessel's contents in a place where they could not recover most of the spilled grains. A month or so later, people who happened to be passing by the place where the spill occurred, were surprised to see a new, thick stand of wheat growing there, where there had been nothing of that sort before. As a result of that incident, the whole tribe might conclude that (depending on the time of year, the amount of rainfall, etc.) they vastly would be able to increase their supply of wheat grains, if they planted—in nearby plots of good soil—grains they already had gathered, and then returned to those same plots later to harvest the new wheat that had grown from the planted seeds. On the other hand, how could any similarly accidental occurrence lead humans to suppose that it somehow would be possible for them to induce or persuade, either carnivorous animals with whom they once had competed for food and living space, or herbivorous animals they previously had hunted and killed as prey, to put aside their natural fear, suspicion, and loathing of

[4] Similar changes also often happen when animals become marooned on small, isolated islands. Some experts claim that this pattern of bodily changes occurs because it is part of the animals' adaptive response to the limited supply of food available on such islands. See for example, Wong, 2005, p.59. But this last idea is a bit puzzling to me, since I do not find it plausible to suppose that domesticated animals also would have access to less food than the supply that is available wild animals.

[5] A simple and clear—but not current—account of these topics appears in Chapter 4 of Leonard *et al.*, 1973. See especially, pp.81-4.

human beings, and to begin to live with them in a peaceful arrangement of coexistence?

All the would-be scientific accounts of the beginning of domestication that are known to me (e.g., the one proposed by Jared Diamond in his book, 1999) assume that the ancient people who first domesticated the plants and animals that have become familiar staples to us today—wheat, barley, rice; dogs, sheep, cows, pigs, horses, etc.—must have thought, reasoned, and motivated themselves in essentially the same ways as farmers, cultivators, and breeders do now. Thus, according to those theorists, the people of the time in question could have solved many of the problems that were connected with their attempts to domesticate animals simply by "talking to themselves." The theorists who propose accounts of this sort assume that the ability to speak in a syntactically organized fashion is something that became centrally important to all undamaged members of our species, immediately after the UPR or Great Leap Forward, and that domestication of animals took place following the UPR. But this is not an idea that I accept. Rather, I consider it more plausible to believe that domestication took place before the occurrence of the UPR, and was one of the factors that prepared the ground for the UPR, and allowed it to take place at a later time.

To repeat: In the rest of this chapter I shall argue in favor of the following two propositions. (1) Humans' domestication of the first of the non-human animals, with which they eventually were to become associated, probably happened before humans were able to speak in something like the same sophisticated and syntactically organized way that we ourselves employ at the present moment. Furthermore, (2) the domestication of those animals acted as a model and guide that helped humans to arrive at the various distinctive forms of thinking—including but not limited to linguistic thinking—that they developed since that time. To be still more specific, instead of assuming in the way Diamond does, that the first humans who domesticated animals had approximately the same level of linguistic competence (and thus competence in thinking) as people do now, I intend to argue that—in the absence of sophisticated language—humans got the idea of how to domesticate animals from observing certain characteristics and behaviors that belonged to those animals themselves.

3.2 Two clues from early hominid history about the background of the nature we now possess: (A) The biological isolation of homo erectus, and (B) The "Pit of Bones" in Spain

In the preceding pages, I nearly always have used "humans" and "human beings" to refer, in a narrow sense of these words, to people like us—i.e. modern members of the species *Homo sapiens*. But a stricter understanding of the terms would extend their reference beyond our species alone, to include quite a few other species as well—each one now presumably extinct—which together with ours, make up the whole primate genus known as the hominids (roughly speaking, primates who are both bipedal, and who also make relatively sophisticated tools). We have no reason to suspect that any of our hominid ancestors or relations ever succeeded in domesticating creatures of other species. Nevertheless, certain things happened to some of those earlier hominids that might have helped to prepare some of their successors (more particularly, us) to become able to do that later. More generally, I now shall try to give more substance to the remarks that ended the previous section, by talking about certain past events that helped sapiens like ourselves to begin thinking and acting in the ways we do at the present time.

In spite of the fact that I am not an expert in the fields of archeology and paleontology, I would like to say a few things in this section, about two relatively early species of hominids. The first of those species is the one to which we now give the name of *Homo erectus*, and the second is a species that was ancestral to the European Neanderthals (whose particular species-name is unknown to me) some of whose bodily remains were discovered in a place now informally known as the "pit of bones" in Spain (or *Sima de los Huesos* in Spanish).

Scholars think that *Homo erectus* occupies a pivotal place in the history of our genus for at least two, presumably connected reasons. The first of those reasons is that this species apparently lived for a much longer time than any of the other species of the hominid genus, including (so far) our own. According to recent estimates, it survived in one or another form for about a million years—and, in view of the fact that the whole genus itself existed for only a bit more than 2 million years, this is roughly half of that time. Furthermore, *Homo erectus* eventually became the only surviving species of the hominids, so that it was necessary for all the new species that constituted the next generation of this genus to arise from it.[6]

What led to this species' becoming biologically isolated in the way I just have described? The apparent answer—as well as the second of the two reasons experts have for supposing that *Homo erectus* has a special place in the history of our genus—is that, roughly two hundred thousand years after it first came into existence, some of its members went through a fairly dramatic change in their thought and behavior. The change to which I now am referring is that those human beings became able to visualize the general form that a stone tool would take, before they began to create that tool. For instance, Ian Tattersall says (1998, pp.138-9):

> A couple of hundred thousand years after fossils of *Homo ergaster* [an early form of *Homo erectus*] show up, . . . we . . . see a remarkable cultural innovation in the archaeological record. Up to that time (about 1.5 myr [million years] ago), stone tools had been of the simple kind that had been made for the previous million years or so, in which the main aim had probably been to achieve a particular attribute (a sharp cutting edge) rather than a specific shape. Suddenly, however, a new kind of tool was on the scene: the Acheulean hand ax and associated tool types, which were obviously made to a standardized pattern that existed in the toolmaker's mind before the toolmaking process began. Hand axes are large, flattish, teardrop-shaped implements that were carefully fashioned on both sides to achieve a symmetrical shape; and because of the multifarious uses to which they were evidently put, they have been described as "the Swiss Army Knife of the Paleolithic."

What factors cause living creatures (or series of such creatures) to change their biological characteristics? As already noted, Charles Darwin claimed the main such factor was the principle of natural selection, or the preservation of favorable characteristics and elimination of unfavorable ones through the selective death of those individual creatures that were relatively less adapted to their environments. Furthermore, Darwin thought that natural selection always worked in

[6] It is possible to infer this point from historical diagrams or "maps" like those that appear, for example, in Leakey, 1994, p.33; Tattersall, 1998, p.185; Wong, 2005, p.65; Zimmer, 2005, p.41; and Wong and Deak, 2009, pp.61-3. However, recent reports of the discovery of 13 thousand year old remains of three-foot high, stone tool making hominids in Flores Island in Indonesia, and of similar, 1.75 million year old creatures that might or might not be of that same species (but which so far have not been found to be accompanied with stone tools) in Dmanisi, Georgia show that the supposed isolation of *Homo erectus* might not have been an absolute one, but instead might have been a relative matter.

a constant and gradual fashion. According to his view, for example, there might have been a group of animals whose characteristics would prompt human observers to refer to them as "lions." But if that group became separated from the great majority of other lions for a long period and if, during that same time, they were subjected to different adaptive conditions from those that applied to most other lions, then that group gradually might have changed in ways that would lead later people to speak of—and identify—them as the different species (or sub-species) of "tigers."[7] In contrast with this, a pair of invertebrate twentieth-century paleontologists, Niles Eldredge and Steven Jay Gould (see 1972), concluded on the basis of their respective investigations of trilobites and land snails, that almost all change in animals was a result of speciation (i.e. different separate species rapidly replacing one another), rather than of single individuals gradually changing their properties through time.[8] Furthermore, in a manner consistent with that idea, Gould claimed that our own species had not changed in any major biological respect for a very long period (at least for the last 50,000 to 40,000 years), so that everything humans had accomplished since that time had been achieved with the help of culture and civilization, instead of originating from bodily changes.

I do not believe that either Darwin's or Gould's view can provide a good explanation of what happened in the case of erectus, since neither of those theories is suited to account for the special kind of change that took place in that case. To be explicit, Darwin's view cannot do this because natural selection only concerns itself with simple survival; and therefore it always operates in ways that are disjointed, opportunistic, and piecemeal, rather than general and systematic. However, the mental transformation of erectus about which we now are speaking was neither gradual nor piecemeal, as

---

[7] We usually think of lions and tigers as separate species. But under certain conditions, they can interbreed and produce healthy offspring. (It is not clear to me whether or not these offspring are, or at any rate tend to be, fertile.) On this subject, see Wilson 1992, p.39. Lions also are able to interbreed with leopards. For instance, on page 16 of an old children's book my grown daughters once read, and which is now being enjoyed by my four year old granddaughter, there is a photograph of a product of such a union—a "leopon"—which possesses both a lion's mane and a leopard's spots. See Editors of the How and Why Library, 1964-76.

[8] The following two sentences occur in Gould's paper, 1980c, p.183: "Eldredge and I believe that speciation is responsible for almost all evolutionary change. Moreover, the way in which it occurs virtually guarantees that sudden appearance and stasis shall dominate the fossil record."

shown by the fact that this transformation became the foundation of a new dimension of behavior and thought that changed our entire genus in an apparently permanent fashion.

Some recent theorists have criticized Gould and like-minded individuals for claiming in an unsupported fashion that the biological evolution of our species ever came to an end. For example, consider some excerpts from a newspaper report by Lynda Hurst (2008) that summarizes the sort of scientific evidence that supports that criticism:[9]

> Mankind's earliest ancestors split from the forerunners of today's chimpanzees about 6 million years ago. Roughly 2 million years ago, the predecessors of modern humans began the long trek out of Africa and into the rest of the world.
>
> About 150,000 years ago, we appeared, modern humans. Some 90,000[10] years later, our brains made a stunning leap forward, developing complex language and abstract symbols. We had begun the journey to civilization.
>
> At that point, the evolutionary process, having sufficiently ensured humans' survival as a species, basically stopped, slowing to a glacial pace. Or so it was thought.
>
> By the late evolutionary biologist Stephen Jay Gould, for instance. In an essay published in 2000, he wrote, "there's been no biological change in humans in 40,000 or 50,000 years. Everything we call culture and civilization we've built with the same body and brain."
>
> Noted British geneticist Steve Jones broadly agreed, but dated the evolutionary slowdown much later, with the rise of agriculture at the end of the Ice Age 10,000 to 12,000 years ago.
>
> When humans made the transition from hunting-gathering to raising crops and domesticating animals, the move led to dietary changes and to settled habitats in specific regions. Combined, they ignited a surge in human numbers.
>
> Far from slowing down, it appears that, when there were enough people to, in effect, work with, the process of evolution rapidly began to accelerate.
>
> Even without modern-day knowledge of genes, Charles Darwin wrote in his revolutionary The Origin of Species that in animal breeding, herd size "is of the highest importance for success" because large populations have more genetic variation. The same turns out to be true for us.
>
> Since the advent of agriculture, the human population has grown steadily from about 5 million in 1 A.D. (it's 6.5 billion today). But as people migrated to different geographic regions, they had to adapt to a variety of conditions and pressures.

---

[9] Also see McAuliffe, 2009.

[10] I have taken the liberty of correcting this number, as it appeared in Hurst's article. (She had written the number: "50,000.")

> One example cited by [a new study by John Hawks *et al.*] is lactase, the gene that helps humans digest milk but which, for most of the planet's population, switches off in adulthood. At some time in the past few thousand years, northern European dairy farmers—living with weaker sunlight therefore less vitamin D exposure—developed a mutation that lets them tolerate health-giving milk throughout their lives.

I do not disagree with general tone of the criticisms Hurst mentions, against the views of people like Gould and Steve Jones. (Also see Ward, 2009.) Nevertheless, criticisms of that sort do not shed light on the question we asked before—namely, why a Gould-like picture of evolution cannot account for changes like the one introduced by erectus. I think the most plausible answer to this question is that accounts like those proposed by Gould and similar thinkers (including Darwin) are not able to make sense of what happened in the case of erectus, because they assume that all changes in animals either are, or are summaries of, detailed changes in those animals' biological characteristics. However, the change in erectus was not just an accidental, cumulative result of smaller changes of that sort. Instead, it was systematic and foundational in its nature because of the fact that it was an expression of, and was motivated by, a general idea.

Some biologists (e.g. Wilson, 1992, p.201) say that all of the niches that exist in nature already are filled. But this idea is obviously wrong, because of cases we previously noted like the long empty bat niche, and the bee-like niche for birds, which ancestors of the humming birds first discovered and then also filled. Thus, I suggest the best way for us to make sense of the systematic change erectus inaugurated is to describe it as the discovery of a new adaptive niche.[11] That is, erectus was the first hominid species—in fact, the first species of any kind—to move into a new, more complex and effective style of life based on the reflective practice of looking, imagining, and planning ahead. Because of the fact that members of *Homo erectus* had adopted a primitive version of that way of thinking and making a living, other existing species of hominids eventually found it impossible to compete with them, with the result that all of the last mentioned species eventually became extinct. To adopt a metaphor from mechanics, we might say that all the other hominid species living at the same time eventually went out of existence because erectus had

---

[11] An alternative way of speaking—advantageous for some purposes—is to describe it as a particularization, deepening, and improvement of the more general niche that earlier hominids already had succeeded in occupying.

"ratcheted up" the standard of what it meant to operate as a competent and successful hominid. Another, still simpler way of describing that new standard is to say that erectus had got the idea of putting an increased amount of "room" between a thinker on one side, and the objects of his or her thoughts on the other.

Let me now change the focus of our discussion, by moving on to a more recent example that, in certain respects, is more specific than the broad, "conceptual" one associated with erectus, with which we were concerned before. Experts say that early hominids mostly lived under the sky and stars, and only occasionally—and for short periods—took refuge in caves. Nevertheless, because caves usually provide a better environment for the preservation of bones and artifacts, than the open air, a large majority of the ancient human remains we now possess have come from caves. In 1976, archeologists working in a cave in Spain were lucky enough to stumble upon an unusually rich collection of human remains that had belonged to some of the forebears of the Neanderthals. In fact, this single place now has yielded more than 2000 humans fossils, in spite of workers only having excavated limited parts of it so far, and to a very shallow depth. The Spanish archeologist, Juan Luis Arsuaga, sums up the scientific significance of this place by saying (2002, p.222) that this archeological site—just by itself—has produced a greater number of fossils than any other site of the genus *Homo*, which are older than the modern human burial sites of the late Upper Paleolithic.

One interesting fact about the thirty-two separate humans so far recognized in the pit of bones is that all of them fall into a comparatively narrow range of ages and physical conditions. That is, almost all the bodies are of individuals in their prime years (very few old people or children are represented there). Also, there are no bodies of people whose bones show that they were deformed, or that they otherwise had been handicapped in one or another way, in life. Some archeologists (including Arsuaga) claim that the most plausible way of explaining this pattern among the remains is to suppose there was one or several ecological crises—for instance, a long period of unusual heat, or a serious drought, or several consecutive years of particularly long and cold winters—which drove those people to come into that area of Spain, on an emergency basis.

Juan Luis Arsuaga proposes to give flesh to this idea (2002, pp.230-2) by saying that although widespread epidemics like the plagues of medieval Europe would not have been possible in the period of Sima de los Huesos, some contagious diseases might have affected small human groups at that time. Nevertheless, the age profile of the thirty-two individuals discovered at Sima de los Huesos show that this last hypothesis did not apply to their case, because the pattern of their deaths was not typical of disease. For example, he mentions two modern and therefore well-documented epidemics, one of cholera and the other of smallpox. In both of those events, most of the dead were under ten years old—forty-five percent in the first case and ninety percent in the second. The presumed reason for this is that epidemic diseases generally kill more young children than adolescents and young adults. Nevertheless, precisely the latter two groups are the ones that predominate in the bones that were found at Sima de los Huesos.

Thus if it really is correct to suppose that some ecological disaster accounts for the pattern of the remains found in the Pit of Bones, the reason for this must lie in the fact that humans differ from other animals, because they are not content to wait passively for a crisis to pass. Instead, these humans presumably moved out of the affected area to search for better conditions. During this mass migration, the weakest members of their group—the children, the aged, the sick and disabled—would have fallen by the wayside, so eventually only the adolescents and young adults were left, because they were the ones that were more physically robust than the others.

We can suppose that, after a difficult journey, those strongest individuals managed to reach this Spanish mountain refuge, leaving many fallen friends and family members along the way. Once they arrived, their suffering continued on for a time, and perhaps many of those individuals arrived in such a debilitated state that they did not succeed in lasting much longer. But in either of those cases, many more of them died. The lucky survivors then looked for an isolated spot in which to deposit the bodies of the individuals who had died after having completed the journey, in order to protect them from the depredations of carrion eaters. They found a large cave with only one, very narrow mouth for ingress, which admitted virtually no light. Both because of its difficult access and its lack of light, the cave never had been occupied by humans before, although bears had used it year after year for hibernation. In one corner of the cave, not far from the entrance, there was a mysterious vertical shaft almost forty-six feet deep, although a person could not see its bottom from above. It was here that they deposited the cadavers of their deceased family members and friends in what was, as far as we can tell from available evidence, the first human funerary activity. The ecological crisis then passed. Animal and human populations recovered; and life continued as it had for ages in the peninsular interior of Spain. But in the Burgos area, there was a cave that contained the remains of at least thirty-two humans, who had lived 300,000 years ago. At some point the entrance to the cave was blocked by natural causes, so that bears no longer were able to enter it for winter hibernation. In fact, nobody at all visited the Sima de los Huesos again until it was rediscovered by investigators in the twentieth century.

I now suggest that we can plausibly think of this second case as still another, in some ways a surprising, expression (a further "ratcheting up") of the same ecological niche erectus previously had found.[12] The central idea of that niche—and therefore also, what I take to be the basic mark that separates modern humans on one side from both non-human creatures and pre-modern humans on the other—is the feature of considering oneself as different and separated from one's environment, and therefore as also having a power to contemplate and work on that environment in some non-immediate, instrumental, and "controlling" fashion. To express the same point in a different fashion, that which now sets humans apart from all the other earthly creatures has been a result of a "trick" that relatively late humans learned, and then passed down to their descendents. The trick about which I now am speaking is that of putting space between themselves and the objects of their thinking, in such a way as to allow them to envisage what is likely to happen (e.g. as a result of an ecological crisis) before it happens.[13]

---

[12] It not surprising that the remains in the Pit of Bones should have come from a period later than the time when *Homo erectus* was virtually the only species of hominid that existed on Earth because, as noted before, the Neanderthals and all of their predecessors presumably were descendents of the members of the species *Homo erectus* who lived during the time of that species' exclusiveness.

[13] Hurst says somewhere in her article that most biologists refuse—presumably for political reasons—to speculate about the future destiny or fate of the human race, as we know it today. Nevertheless, in view of the fact that our own species also has become biologically isolated in its genus in approximately the same fashion as erectus once was, it seems plausible to believe the previous career of erectus might have been a "trial run" for the kind of future that our species eventually will face as well. In other words, in my admittedly non-professional opinion, the increased rate of evolutionary mutation our species now is experiencing eventually might lead to a whole new suite of hominid species coming to exist at some future time, out of present-day *Homo sapiens*.

3.3 Entrapment vs. attraction: What was it necessary for the first domestic animals to be like, in order for them to "Tame Themselves"?

> So they wake: first the catbirds and cardinals and some I do not know. Later the song sparrows and the wrens. Last of all the doves and crows. The waking of crows is most like the waking of men: querulous, noisy, and raw.
>
> Thomas Merton

Having discussed introductory matters in the preceding two sections, we now are ready to come to grips with this chapter's main concern—namely, a historical consideration of the domestication of animals. There are at least two possible approaches investigators can take with respect to this subject. The first, currently more popular one is to ask what particular characteristics certain animals must have had, in order for them to be domesticated by humans. The second, less familiar approach is to ask what changes must have taken place in humans themselves, for them to become able to domesticate other animals. I shall begin by talking mostly about the first of these approaches in this section, and then switch, in the chapter's next and last section, to a consideration of the second.

A quasi-popular book that fell into my hands many years ago—the one that first got me interested in the general subject of domestication (Leonard *et al.*, 1973)—posed the question of what the first species of animals was that humans had domesticated. The authors of that book had a strong desire to follow what they conceived of as a strictly scientific procedure. For that reason, they based their proposed answer on the oldest remains (teeth and bones) of various candidate animals, which were known at the time the book was written. This odd procedure led them to accept the idea that the sheep had been the first domesticated animal, and the dog had been the second animal that humans domesticated after the sheep, simply because of the fact that the then oldest known sheep remains (found at Zawi Chemi Chanidar in Iraq) were dated at 8500 B.C., while the oldest dog remains (found at Jaguar Cave in Idaho) were dated at 8400 B.C. (Ibid., p.77.)[14]

---

[14] I do not propose to talk about the case of sheep in this chapter. But archeologists have uncovered still older remains of dogs that were associated with human beings, since the time of the publication of that book. For example, in an article written by

In retrospect, I do not consider that procedure to have been either sensible or sound. That point becomes clear when one reflects on the facts that (a) preserved bones more than 10,000 years old are quite scarce, and finding them is largely a matter of luck; furthermore (b) in addition to evidence associated with bodily remains, we also have a great deal of indirect but convincing evidence that supports the idea that dogs were domesticated a very long time before sheep were. As an example of evidence of this last sort, consider the following short excerpt from a popular article on the general subject of dogs (Phillips, 2002, p.15):

> Roddy MacDiarmid, 57, lifelong shepherd and son of a shepherd, surveys the Scottish Highlands from a ridge overlooking Loch Fyne and the little valley town of Cairndow. On one hand lies the estate of John Noble, where MacDiarmid has worked much of his life, on the other the estate of the Duke of Argyll. Black-faced lambs and ewes by the hundreds dot the green hill-sides below. His Border collies, Mirk and Dot, trot faithfully behind. It's familiar turf. "Everywhere you see," says MacDiarmid, sweeping his shepherd's crook in an all-encompassing arc, "I have gathered sheep. And I can tell you this: You cannot gather sheep from these hills without dogs. Never could and never will; never, never, ever!"

Another piece of indirect evidence that points to the conclusion that dogs must have been domesticated at least by 60,000 years ago is the presence today of what seem to be "re-wilded" dogs (popularly known as "dingos") in Australia. Dingos are placental mammals of the kind that are native to the Old World (Africa and Eurasia). They are not marsupial mammals of the sort that developed in Australia. Thus, how did they succeed in getting to that isolated continent? Most experts believe the most plausible hypothesis about how that happened is that their ancestors were domesticated dogs that arrived on rafts from Asia, along with Australia's first human settlers, at the very early date of approximately 60,000 years ago.[15]

Karen Lange (2002, p.4) a picture appears of the excavated bones of a puppy buried in a human grave in Israel about 12,000 years ago.

[15] An article written by Nicholas Wade recently appeared in my newspaper (*The Globe and Mail*, Thursday, March 18, 2010, p.A3) bearing the title: "From the big, bad wolf comes Fido and Fluffy: Researchers analyzed thousands of genetic markers to find that today's pet dogs descend from Middle Eastern grey wolves." As a further means of summarizing the message of this article, three photographs of canines were placed just above the title. As a caption for the first photograph (of a wolf) was the sentence, "It all started 20,000 years ago with a wolf." As a caption for the second photograph (of an Australian dingo) was the sentence "Somewhere in the middle

Dogs are not now, and only rarely have been in the past, primary sources of food for humans. Instead of sources of meat, it was more natural for humans to employ dogs as guard animals, hunting companions, and pets. Only later, once the Neolithic era had begun, did dogs also become helpers for herders and farmers. In any case—taking account of the preceding points—I shall assume in what remains of this chapter that dogs (wolves), rather than sheep, or animals of any other type, were the first animals humans domesticated.

In general, what it means to tame an animal is simply to calm its fears and make it willing to live peacefully, in close proximity with humans. To domesticate an animal is to do all of that and, in addition, also to control its food, living conditions, and breeding, so as eventually to influence its genetic make-up. (See Diamond, 1999, p.159.) Thus, let us pose two questions. (i) What encouraged and enabled humans to take the first step of taming dogs? (ii) How and why did humans then go on to take the second step of domesticating those animals as well?

Karen Lange says (*op. cit.*, p.4) scientists are aware of the fact that the process of domesticating dogs from wolves was under way at least by 14,000 years ago, but they do not agree on exactly why and how this took place. Some of them argue that humans adopted wolf pups as pets, and natural selection then favored those puppies that happened to be less aggressive and better at begging for food than their fellows were. Other scientists claim that dogs domesticated themselves by adapting to a new niche—namely, the niche of scavenging food that they were able to find in human refuse dumps. The scavenging canids that were less likely to flee from people survived in that niche, and succeeding generations of them became increasingly tame. She quotes the biologist Raymond Coppinger as saying that the one necessary trait that was selected for in this process was simply the

---

came the dingo." And as a caption for the third photograph (of a domestic dog) was the phrase "Arriving at this modern-day Chihuahua."

I cannot claim to know a great deal about genetics. Nevertheless, I think it is clear that the numbers mentioned in these captions do not correspond to facts archeologists have discovered about human prehistory. In particular, half (the "middle") of the number 20,000 is 10,000. But archeologists (perhaps as opposed to some geneticists) definitely do not believe that the dingo came into existence 10,000 years before the present. Instead, a great deal of archeological evidence shows that dingos (or at least their immediate ancestors) already existed at least by the date of 60,000 years ago, when Australia first was colonized by humans.

ability to eat in close proximity to people. She then adds that not much changed at all at the molecular level when dogs were domesticated, because the DNA makeup of wolves and dogs is almost identical.

Of the two alternatives just mentioned, the first one (adopting wolf cubs) is not a sensible or adequate answer, since it presupposes precisely the point that needs to be explained. That is, this alternative assumes that some of our human ancestors somehow got the new idea of "mothering" members of another species, by persuading those creatures to live with humans in close and affectionate circumstances (which had not been the case before, either with any other humans or with any other wolves); but it does not say how the people in question succeeded in getting the idea of doing those things in the first place. The second alternative (wolves domesticated themselves) strikes me as more plausible. Admittedly, however, there still is at least one problem with it. The problem is this: As we noted before, along with our ancestors, relatively advanced hominids of two other species also lived in the same areas at the same time (Neanderthals in Europe and the Middle East, and a late version of Erectus in East Asia). It presumably is correct to say that those other hominids also created refuse heaps of their own. Furthermore, those heaps also must have attracted scavenging wolves, in the same manner as the garbage associated with members of our species did. Why did *Homo sapiens* manage to domesticate wolves/dogs, when neither the Neanderthals nor Erectus were able to do this?

In his book, *Guns, Germs, and Steel* (1999, pp.168-73), Jared Diamond lists six properties an animal must have in order for that animal to be "domesticable." These are (i) it must not be too finicky in its food preferences to recommend itself as a farm animal; (ii) it must grow quickly enough to make it useful for farmers to raise it; (iii) it must have the ability to breed in captivity; (iv) it must not be physically dangerous or have a nasty disposition; (v) it must not have a tendency to panic; and (vi) it must have a hierarchical social structure that humans can exploit in order to assume a position of authority over it, by taking over the position that once was occupied by the wild "alpha animal." However, we must put the following question to Diamond: Since it presumably is true that the wolves that came to the garbage heaps of the Neanderthals and of *Homo erectus* had the same characteristics as those that came to the garbage heaps created by homo sapiens, why did only members of the second group become domesticated? Diamond must have left out something important from his account of domestication. I suggest that what he omitted was the fact that, in addition to the above list of six properties, a domesticable species also needs to have a certain special attitude towards humans.

People sometimes speculate that the Neanderthals probably conceived of themselves—in some respects—as different from all other creatures, as shown by the facts that they were the first hominids to wear clothes, that they also buried their dead, and that they also gave help and care to the sick and disabled members of their group. I grant that there is something correct about that idea. That is, the behaviors just mentioned were a sign that Neanderthal humans had traveled part of the distance down a road that eventually might have given them a power to domesticate animals. Nevertheless, the Neanderthals did not succeed in thinking of themselves as different *enough* from the rest of living things, for them to develop a power to domesticate some of their fellow creatures. By contrast, some of our own ancestors were able to invent the revolutionary practice of domesticating other species, because they had traveled further along that same, erectus-initiated path of separating and dissociating themselves from their environment, than any of the other species of hominids.

There are several possible (non-exclusive) ways in which Neanderthals and late members of the species *Homo erectus* might have reacted to wolves. First, the humans might have had a fear of wolves that they never were able to conquer. Or, second, they might have felt a certain sort of anger and irritation towards those animals that they never learned to overcome. Third, they might have had a haughty distain for wolves that never disappeared. By contrast, let us imagine a situation in which, even though our forebears started out with one or more of those attitudes at first, their attitudes towards wolves changed at a later time (at least in the case of some of the sapiens), and was replaced by something more like curiosity and compassion. The wolves then noticed this tolerance among sapiens, and decided to take advantage of it. In other words, they now began to see sapiens as presenting them with a new opportunity they had not encountered before; and then they set out to exploit that opportunity.

I reject the idea that wolves are, or that they ever were in the past, "natural slaves" (to use Aristotle's term) that were ripe and ready to accept human exploitation. Rather, they are now, and probably also have been for many millennia, creatures that are fearful, suspicious, self-sufficient, and proudly independent. Accordingly, different groups of wolves might have developed both of two opposite responses to early humans—some of them continuing to conceive of themselves as in competition with humans, and others beginning to tolerate, to work with, and to exploit humans for their own advantage. At the beginning, relatively weaker, less adjusted, or outcast wolves might have been the ones that were more likely to become domesticates of humans than those that were stronger, more confident, and more self-reliant. But even if that situation existed at first, things probably changed later. For example, I read somewhere that farmers in Northern Ontario tried many different strategies to protect their herds of sheep from wolves, but nothing worked until they finally lit upon the idea of "fighting fire with fire." In other words, they learned—after a long, slow, and painful search—that the only effective response to the wolves' depredations was to challenge them with domesticated sheepdogs.[16]

Historically speaking, one of the ways just mentioned, in which wolves related themselves to humans, turned out to be far more successful than the other. At the present time, wolves (in the form of dogs) have become—like chickens—one of the most numerous, diverse, widespread, and securely established species on the planet. Contrast the response these animals have to humans, with the attitude towards humans typically taken by crows. Crows also are a successful, resourceful, and intelligent species. But neither humans on their side, nor crows on theirs, ever have been able to find (or accept) any realistic means or methods by which their two species might learn to cooperate for their mutual benefit. Thus year after year—especially in winter, when the crows congregate in enormous flocks or "roosts"—the rancorous, wasteful, and ultimately fruitless battle between these two, tragically mismatched species continues to play itself out.

Consider the following short article that appeared in my newspaper (Dube, 2007):

> Every day, the city of Chatham collects garbage from a different neighbourhood. And somehow, every day, the crows know exactly where to go.
>
> They wait patiently on rooftops and tree branches as the sun rises. When the trash goes out, the carnage begins. Crows descend upon bulging plastic bags, disembowel them in search of food, and gleefully scatter the contents across front lawns.

---

[16] Can people in Southern Ontario (like me in Toronto) use a similar technique to keep raccoons from ravishing our green garbage cans filled with food scraps? The answer is no, because raccoons have not shown any signs of being domesticable.

Professional crow-chaser Ulrich Watermann just laughs and shakes his head as he watches the birds.

"People," he says, "are incredibly stupid."

Crows, on the other hand, are smart—smart enough to memorize garbage routes and a lot more. Humans, with our big brains and opposable thumbs, should theoretically be able to outwit them. Yet fat crows and messy lawns remain common sights, not just in this southern Ontario town, but in crow-plagued cities from Burnaby, B.C. to Charlottetown. . . .

Humans have feared crows as evil omens and reviled them as urban nuisances, but only now are we starting to understand just how intelligent they are. Their cleverness in the lab and in the wild has made them scientific darlings and YouTube stars. But the cunning that fascinates researchers also stymies towns that are besieged by thousands of noisy, messy crows every winter.

The battle between birds and humans has spawned a cottage industry of professionals such as Mr. Watermann, whose bag of tricks includes pyrotechnics, lasers, recorded distress calls, birds of prey and wooden noisemakers.

He constantly changes his mode of attack, hoping to annoy the crows so much that they leave.

If the birds flee to a neighbouring town, he'll be happy to help them too—for a fee. Chatham pays Mr. Watermann about $32,000 a year.

Lean and dressed in black, with glittering blue eyes and a weathered face, Mr. Watermann could pass for an overgrown crow himself. Like most people who study crows, he's grown to admire them.

When hunters fired shotguns at the Chatham crows, he says, the birds figured out exactly how high to fly to escape the pellet range. When city workers made nightly rounds to disturb them, the crows learned that the workers clocked out at 11 p.m., and simply waited until 11:01 to head into town for the night.

"They are highly intelligent as a single bird. Then they come by the hundreds of thousands, and they all learn from each other. If that one makes a mistake," he says, pointing to a crow regarding him suspiciously from a wire, "the next one doesn't make it."

In labs, crows have demonstrated that they can not only use tools, they can make tools—bending a straight wire into a hook to pick up a basket of food, for instance. They will also use one tool to fetch another tool for a specific task, a skill known as "meta tool use." With special "crow cams," researchers have observed similar feats in the wild. Experts say crows rival primates in intelligence.

Smart crows also make great video. One BBC clip on YouTube shows crows dropping nuts onto a busy road to be run over and cracked open. Then they wait for the crosswalk signal to fly down and retrieve their treats in safety.

One YouTube commenter wrote in response: "If they had hands like us humens [sic], they would take over the world and enslave us!"[17]

Residents of towns where they roost may be forgiven for thinking the crow revolution has already arrived. Mr. Watermann believes the Chatham roost once numbered in the millions. (Current estimates range from tens of thousands to 200,000.)

At the peak of their infestation, crows pooped on houses, churches, businesses, cars and sidewalks. They wriggled into finished engine blocks at the local truck-assembly plant, leaving brand-new engines coated in guano and feathers. Residents carried umbrellas on sunny days.

"I hate crows," says Scott Heuvelmans, owner of a Chevrolet-Cadillac dealership in Chatham, where they used to wash every car on the lot daily. Now they're down to weekly washes, but he'd still be happy to see the last of the crows.

"I wish somebody would just go out and shoot them," Mr. Heuvelmans says.

The piercing cries and blackened skies as crows descend en masse inevitably trigger comparisons to Alfred Hitchcock's *The Birds*.

"I hate that movie," says Kevin McGowan, a Cornell University ornithologist and crow expert. A winter roost—or to use the poetic term, a murder of crows—"is not spooky, it's not unnatural; it's the most natural thing in the world," Dr. McGowan insists. "They've been doing this for millions of years."

Indeed, crows have always flocked together on winter nights for warmth and protection from predators—usually a few sentinels keep watch while the rest sleep or socialize. Some roosts date back centuries.

But it wasn't until the 1980's that people started noticing crows roosting in North American cities, Dr. McGown says. He's got theories but no answers as to why. Crows like the heat of cities and the streetlights—they can't see well at night, so the light gives them an edge over owls and other predators. The garbage in town is a nice perk, Dr. McGowan says, but he believes most crows do their serious eating in fields.

Chatham in particular attracts crows in winter for the same reason Florida attracts snowbirds: location. The city of about 45,000 is sandwiched between lakes Huron and Erie, at the convergence of two major migration routes. The river running through the center of town provides warmth, and fields of cash crops nearby offer plentiful food. You'd be hard-pressed to

---

[17] A related point recently was mentioned in a short, anonymously authored article quoted from Reuters and BBC News (*The Globe and Mail*, Thursday, August 21, 2008, p.L6). According that article, German scientists have discovered that magpies (another corvid species closely related to crows) can recognize themselves in a mirror. For example, one test that researchers from Goethe University in Frankfort employed to show this was by placing yellow and red stickers on five magpies, in positions where the stickers only could be seen in a mirror. They found that, whenever the birds saw this, they tried to remove the stickers with their claws and beaks. The article's author expresses the opinion that this is the first time the power of self-recognition has been observed in any non-mammal.

> design a better crow haven. Except, of course, for the crow-disturber with his pyrotechnics and trained hawks. Mr. Watermann realizes his methods may seem uncouth, especially when one of his hawks kills a crow instead of just scaring off the flock. But he figures a few dead crows help save thousands that would otherwise be shot or poisoned.
>
> Like Chatham, other cities are turning away from killing crows. That's partly because lethal methods often don't work—Chatham was overrun by so many crows that even the most trigger-happy hunters didn't dent their numbers, and poisoning may kill other wildlife indiscriminately. But it's also because the *corvus* fan club is growing.
>
> "It's not just their intelligence, it's their attitude," says Laurie Ulrich Fuller, a graphic designer in Lancaster, Penn., explaining why the crow became her favourite animal after 60,000 of the birds invaded her town two years ago. "They saunter when they walk; they don't pip along like a little bird."
>
> Lancaster, 588 kilometres southwest of Chatham as the you-know-what flies, uses harassment tactics similar to Mr. Watermann's to disperse crows.
>
> As a founding member of the Lancaster Crow Coalition, Ms. Fuller struggles daily to change public opinion. She shows the BBC video to local groups, which she thinks opens some minds.
>
> But even bird watchers can be shallow, disdaining wisdom in favour of beauty.
>
> They're not pretty birds. Crows don't even get the attention of bird lovers. They'll go ape over a goldfinch, but not a crow," she sighs. "Which is a shame because they're far more interesting and far more intelligent."

As said before, Diamond and like-minded theorists have not taken account of the fact that an attitude some types of non-human animals take towards humans, and which other species do not have, is also an important factor in determining whether any particular species of animals can or cannot be domesticated. How should one describe the attitude that opens the door to domestication? A name we might give it is this: "Tolerance of, and willingness towards making arrangements of reciprocity with, creatures of other species."

In my opinion, the basic reason crows cannot be domesticated is that they are, by nature, far more interested in competing with humans, on humans' own level, than in accepting a subservient, but nevertheless complementary and mutually useful position with respect to them. The last point I want make in this section of the chapter is that—with respect to the topic of domestication—humans themselves are more similar to crows than they are like wolves and dogs, because they tend to think about matters of this sort in a crow-like fashion. It is a familiar fact that, over the centuries, people repeatedly have tried to "domesticate" various groups of their fellow species members, for their own purposes, as opposed to the purposes of the would-be domesticated humans themselves. For example, the ancient institution of slavery is a prominent expression of such attempts. But it gradually became clear to everyone involved in such matters that the whole idea of humans being domesticated by other humans was one that—in addition to being immoral and unjust—was unstable and not sustainable. The suggestion I want to make here is that the social arrangement of slavery is something that neither the supposed slaves nor the attempted slave-masters found to be natural, because it was in conflict with the basic human nature that all of us share.

3.4 What changes had to occur in humans' cultural life, before the domestication of animals could take place?

It sometimes is true that the order in which things happen is important. For instance, Shakespeare scholars point out that three people came into the tomb of the drugged and sleeping Juliet on the first night after her supposedly dead body had been put there—a priest who knew about the drug, and her cousin Tybalt and her husband Romeo, neither of whom knew about it. But if those three people had arrived in a different order, then things would have turned out well for the star-crossed lovers, and *Romeo and Juliet* would have counted as one of Shakespeare's comedies rather than as one of his tragedies.

A point already mentioned several times in this book, and which I shall explain in still more detail in the next chapter, is that many anthropologists believe the advent of syntactical language was that which initiated the advance in human thinking and behavior to which scholars now refer as the Upper Paleolithic Revolution (UPR). As part of that same point, is it appropriate to suppose that our ancestors only gained the power to tame and domesticate animals, shortly after the time (approximately 60,000 to 30,000 years ago) when they also obtained a relatively advanced ability to speak? I believe the answer is no because, instead of our ancestors' ability to domesticate animals being just one more part of that same intellectual advance, I think it was necessary for them to have acquired certain basic aspects of the ability to domesticate animals before they also could develop the power to enter the UPR. To say the same thing in a more direct fashion, domestication already had to be in place before humans could go on to acquire sophisticated speech as well.

Why do I accept this idea? As a means of answering that question, I invite readers to consider the following hypothetical story. Imagine that, more than 60,000 years before the present, there was a group of non-language-using sapiens, whose garbage piles attracted wolves. We can assume that most of the members of this group were sensible enough to stay as far away as they could from those powerful and unpredictable scavengers. Nevertheless, some of them—who happened to be a bit braver than the rest—may have been inclined to try to drive the wolves away, to lessen the danger the nearby presence of wolves posed to humans of the tribe. Some others, who were braver still, might have tried to kill some of the wolves, in order to discourage other wolves from coming near their settlement. Finally, there also appeared courage of a different type, which was shown only by a very small number of young and foolish members of the band. These last mentioned people formed a desire to engage in the daredevil activity of facing down wolves at close range.[18] However, when they at last got up the nerve to do this, they were surprised to find (as noted in the previous section) that some of the wolves were less shy and less aggressive than they had expected them to be. Thus, they gradually became more relaxed around the wolves, and eventually did not bother to bring along weapons whenever they interacted with those creatures. Finally, one or two of the sapiens succeeded in feeding a few of the wolves by hand, and also managed to touch them.

Wolves are clever and resourceful animals with a talent for recognizing opportunities. In my imagined story, the wolves saw such an opportunity in the behavior of the sapiens individuals just described, which they had not seen in the Neanderthals or, for that matter, in any other group of sapiens. This opportunity was that the hominids' flexibility of thought and behavior opened up the possibility that the wolves could occupy a place of mutual advantage alongside those humans.

The humans in our story did not act the way they did because they were prompted to do so by the nature they had inherited from their ancestors. In other words, none of their newly acquired ways of acting was natural in the biological sense of this word. Instead, what they did was based on a once dormant, but now actively exploited ability to put aside many of their inherited inclinations, in favor of an artificial, "second nature" that they fashioned for themselves.

---

[18] This is similar to a kind of behavior one often observes in crows.

The Belgian anthropologist, Claude Lévi-Strauss, once remarked that primitive humans not only considered certain animals and plants good to eat ("*bonnes* à *manger*"), but also good to think with ("*bonnes à penser*")—see Wiseman, 1997, p.54. He rejected the utilitarian notion affirmed by earlier anthropologists like Bronislaw Malinowski, that the thinking of all humans living in relatively non-developed societies was dominated by their basic needs of life so that, for example, they only had knowledge about the species of plants they had found to be edible. Instead, Lévi-Strauss claimed that human beings everywhere and at all times thought about the world in fundamentally the same way, since all of them, including those who lived in undeveloped and pre-scientific communities, had a desire to obtain disinterested, objective knowledge about the objects and creatures around them, for the sake of that knowledge itself. Admittedly, Lévi-Strauss believed primitive people had a different ways of obtaining such objective information about the world, from the means employed by members of more developed societies. Primitives regularly thought about themselves and the social groups to which they belonged, in terms of totem animals, plants and, more rarely, in terms of totem natural phenomena like lightning (see ibid., p.36). But according to him, this totemism was not merely mythical and mystical in character. Instead, it was a practical and relatively objective means that the people in question used to deal with their world. For example, suppose the totem animal assigned to one clan of primitive humans was the jaguar, and the animal assigned to another such clan was the bear. In that case, Lévi-Strauss said, careful examination would show that those choices were neither arbitrary nor misleading, because there were important respects in which the relation that existed between the clans was parallel to the relation between jaguars and bears. Thus, even if both of the clans were composed of hunters, they (like bears and jaguars) did not—and perhaps consciously and purposefully did not—directly compete with one another for prey (see ibid., p.46).

I do not agree with everything Lévi-Strauss has said about these matters.[19] But in spite of that, the general point I want to make is at least similar to his views. It is that once certain animals had become domesticated, it became necessary for humans to think about (and think with) them in a way that was systematically different from the

[19] For example, I do not agree with Lévi-Strauss's claim that humans only gained the power to domesticate animals, after they had obtained the use of language. See Wiseman, 1997, p.66.

way they thought about wild animals. Humans' conception of wild animals was based on instincts they had inherited from their ancestors. But the same was not true of their conception of domesticated creatures. To express this idea more concretely, the people in our story started out feeling the same fears, suspicion, haughtiness, and defensiveness towards wild wolves as their own mammal, primate, and hominid forebears had felt. But when some of the wolves became domesticated (i.e. became dogs), they learned to suspend those natural reactions towards the animals and to replace them with other, learned reactions. In summary, I propose that tamed animals became non-instinctual objects of humans' attention.

How did our ancestors' experience of domesticating animals prepare them to acquire language later? One way it did so was this: Domesticating dogs made them familiar with the activity and process of controlling and re-channeling some of their inborn reactions to items they saw, heard, and felt. Then, when words finally became available to them, it again proved necessary for them to do something similar—i.e. they first had to suspend, then also transform, many of their inherited mammalian instincts, in order for them to make sense of (and employ) those words. Still another reason the experience of domesticating animals might have prepared humans for their later acquisition of language is that close association with those animals in the context of tracking, stalking, hunting, and killing game might have encouraged humans to organize their own thinking (as well as their own society) in ways that were similar to that of a pack of wolves. For example, one result of their association with wolves might have been to make human working groups more disciplined, and organized in a more strictly hierarchical way, than they had been before. It is a basic principle of linguistics that everything that plays a role in a syntactically organized language needs to have a hierarchical form, since language cannot "work" in the absence of this sort of structure. I suggest that the way humans first acquired this specific type of mental organization was from copying the behavior of the first species of animals they domesticated.

Chapter 4

Something else that influenced us: Sophisticated Language conceived as Invented rather than completely Innate, Socio-Cultural as well as Biological

> Language is, of all our mental capacities, the deepest below the threshold of our awareness, the least accessible to the rationalizing mind. We can hardly recall a time when we were without it, still less how we came by it. When we could first frame a thought, it was there. It is like a sheet of transparent glass through which every conceivable object in the world seems clearly visible to us. We find it hard to believe that if the sheet were removed, those objects and that world would no longer exist in the way that we have come to know them.
>
> Derek Bickerton

4.1 How did humans become able to speak?

> Speech is civilization itself. The word, even the most contradictory word, preserves contact—it is silence which isolates.
>
> Thomas Mann

Psychologist and best-selling author Steven Pinker once remarked that he never had met any person who was not interested in the subject of language. This statement is not surprising to me, in view of the fact that we usually consider our command of language to be the one characteristic that most clearly and obviously sets us apart from all other animals. For example, it is apparently true to say that even though some non-human creatures convey simple messages to each other by means of calls, gestures, pheromones and so on, none of them has an ability to communicate by speech. Some small children and child-like adults find it puzzling that other animals lack a power to speak, either with one other or with us. But reflective adults know that humans like us, rather than other creatures, are the ones who are mysterious. In particular, our very existence poses the question of how it ever became possible for us to acquire language in the first place. For instance, some theorists (including Pinker and Chomsky) have been attracted by the idea that the bodies and brains with which we were born were sufficient to produce language as an ordinary biological product, in much the same way as our lungs and brains were sufficient to produce breathing immediately after our births, and our legs and brains were sufficient—at about the age of 1 year—to produce walking. But, for several reasons, it does not seem to me that this idea is correct.

Consider a simple example. I sometimes watch old movies on my computer as a means of relaxing before going to bed. One film I like to see—entitled "Quest for Fire," released in 1982—was shot on location in Scotland, Iceland, Canada, and Kenya. Although it was made for popular entertainment, and was not intended to be a scientific documentary, its makers evidently took some pains to try to portray things accurately. For instance, a publicist claims on the back of the DVD jacket that this movie ". . . offers fascinating insights into prehistoric man's survival." Furthermore, the credits of the film list two contributions made by professional experts—namely, "Special Languages Created by Anthony Burgess," and "Body Language and Gestures by Desmond Morris."

I am not a film critic, any more than an anthropologist or archeologist. Nevertheless, since the time this movie was made, I have learned a few things about discoveries relevant to the topic of human prehistory, by reading books and articles written by scientists. Part of this reading has convinced me that at least one idea of this film must be mistaken. Referring again to the back of the DVD jacket, the movie is supposed to be set in several unspecified places in the Old World, at the time of 80,000 years ago. More specifically, it ". . . follows the lives of four tribes of early man—each with their own customs and stages of development." The problem with this is that each of those tribes is shown as speaking its own complex, fluent, and syntactically organized language. However, (as I already have pointed out in preceding chapters of this book and in the first chapter of the book I published in 2003), we have strong evidence to believe that no language of that type existed on our planet until the later date of approximately 60,000 years ago—i.e. the time of the beginning of the Upper Paleolithic Revolution.

Perhaps the reason the movie proposes an incorrect date for the appearance of language is that the individuals who made the film assumed (in a Chomsky-like fashion) that, for all the members of our species, it never was necessary for language to come into existence at all. What this means is that the moviemakers might have believed that syntactical language always has been associated with human beings of our general type, for just as long a time as humans of that type existed, because a language of this sort was generated in a more or less automatic manner from the special sort of brain and body that all undamaged humans share. For example, they might have supposed that all young homo sapiens children—both now, and also at all times in the past—were "programmed" to experiment with imitating sounds, in such a way as to allow them to absorb any language they happened to hear being spoken around them. (Alternatively, in those very unusual cases in which young children do not hear any language sounds, they assumed that these children had an ability to collaborate with one other in such a way as to invent a wholly new language that never had been spoken before.) Unfortunately for this simple and optimistic conception of humans, however, archeological discoveries have shown that it cannot be correct, because the fundamental nature that belongs to humans (and their babies) has changed over time. To be more precise, unlike the cases of breathing and walking, even though syntactical language comes into existence in a more or less automatic way in the case of present-day human children, this could not also have been true for all the children who were born during at least the first half of the time that our species has existed.

In section 1.3, I discussed anthropologist Paul Mellar's list of seven changes that took place at the time of the UPR (something I also quoted in two of my previous books). I said in that same section that Mellars had left out of account two additional changes—namely, he did not mention that this was when humans started to build ocean-going boats, which they then used for colonizing expeditions, and he did not say that this was when the number of humans increased in a dramatic fashion. (On the second point see Stringer and McKie, 1996, pp.196-7). I now want to add three more considerations that were not mentioned by Mellars. The first of those additional points is that the societies that came into existence after the UPR were larger than any that had been known before. Second, those societies differed from one other in many significant and striking ways.[1],[2] Finally, the third

[1] Klein with Edgar (2002, pp.187-9) say:

> Particular types [of artifacts] are often restricted to certain times and places, which has allowed archeologists to define multiple Upper Paleolithic cultures. Among the most famous are the Aurignacian Culture, which stretched from Bulgaria to Spain between about 37,000 and 29,000 years ago, the Gravettian Culture, which extended from Portugal across southern and central Europe to European Russia between roughly 28,000 and 21,000 years ago, the Solutrean Culture which existed in France and Spain between about

additional point about which Mellars remained silent is that this was when, in the opinion of many investigators, humans first began to speak and to think in terms of a sophisticated, syntactical language. This third point is more fundamental than all the other things Mellars mentions, because (according to people like Ian Tattersall) humans' learning to use a syntactically organized language was the basic cause of the whole suite of interrelated phenomena that suddenly appeared at the beginning of the UPR. (See e.g., Ian Tattersall, 1998, p.232 and 2000, p.62, and Jared Diamond, 1992/2006, p. 364.) In particular, language was the "trigger" that allowed our ancestors to engage in the new type of thinking that quickened the pace and insightfulness of their inventiveness, and helped them meet the challenges and solve the problems that were implicit in all of those new behaviors.

On reflection, I do not think Tattersall and Diamond have probed as deeply into the topic of human nature as they should have done, or as deeply as I am trying to do in this book, because they merely replace the original question of what caused the Upper Paleolithic Revolution with the equally mysterious puzzle of what caused syntactical language to appear at the times and places it did. Thus, they have not really succeeded in explaining and solving the first problem, but merely have re-described the above-listed empirical points in slightly more detail, and from a slightly different angle. To say the same thing another way, the Upper Paleolithic Revolution—along with the special sort of language that played an important role in it—was not magic. It came from somewhere and out of something. But it is not possible to give a satisfactory explanation of what that something was, just by saying that the UPR was triggered by the cultural stimulus of humans' having acquired an advanced form of language, because that idea immediately leads one to ask the exactly

---

21,000 and 16,500 years ago, and the Magdalenian Culture, which occupied France, northern Spain, Switzerland, Germany, Belgium, and southern Britain between about 16,500 and 11,000 years ago.

[2] Some authors (including me) occasionally have exaggerated the novelty of at least some aspects of the UPR. For example, we now know that some human beings made use of ocher pigments (presumably for bodily adornment) as early as the time of *Homo erectus*, and therefore long before the occurrence of the UPR. All things considered, however, it still does not seem to me that such errors of relative detail undercut the general point that needs to be made about the historical importance of this event. (On this subject, also see David Lewis-Williams and David Pearce, 2003, pp.17-8.)

similar question of what created language. Thus, my project in this chapter (in connection with the chapter that preceded it, and the one that will follow it) is to outline a defensible solution to the problem of what the cultural factors and conditions were out of which the special kind of language that brought the Upper Paleolithic era into existence first arose.

4.2 A preparatory comment: To say that certain humans invented language is not to claim (nor does it entail) that those same people also created everything language either includes or presupposes

What does it mean to invent something? To say that person **P** invented device **D** does not imply that **P** brought **D** into existence out of empty space. Instead it means that **P** assembled **D** out of previously existing raw materials. For example, when Alexander Graham Bell invented the telephone, he did that by putting together in a single working assembly, quite a few things that already were known and familiar, like wires, electrical currents, and diaphragms sensitive to sound. So similarly, I claim that if certain human beings invented syntactical language at some point in their history, they did it by fashioning it out of raw (mental) materials that already existed.

What kind of items count as the raw materials out of which someone managed to invent language? Let me propose a single simple case. Noam Chomsky tells us (2000, pp.24-5) that—for unknown reasons—every language linguists so far have discovered and analyzed employs the device of "traces" that presumably act as some sort of means by which the mind (unconsciously) keeps track of hierarchically organized levels of reference. For example, there is a silent trace present in the sentence "Jacks wants to marry Belinda," immediately after the word "wants." Although that trace virtually never is filled in, and only very rarely is recognized in an explicit fashion, one could do that simply by adding the word "Jack" in the place just mentioned, so the completed sentence would read: "Jack wants [Jack] to marry Belinda." A slightly more complicated instance of the same phenomenon, which Chomsky once proposed, is this: "The book seems [the book] to have been stolen [the book]" (see Chomsky, 2000, p.24).

None of us knows exactly what reasons there are for the existence of this particular method or mechanism for organizing sentences. One sign that this is true is the fact that we do not (and probably could not) employ a similar trace system, as a means of helping to organize the various artificial languages we devise for scientific purposes, like the symbolic languages of logic, or the codes with which we program computers. In fact, we do not have any conscious control over our use of traces in language, since we are simply determined to employ them in the way we do, by certain parts of our inherited biological nature. I suggest that the particular function or aspect of our biological make-up that I just have been describing is a simple instance, among many others that might have been mentioned instead, of an inherited resource (i.e. a presupposed bit of raw material) out of which some of our ancestors first fashioned language.

It is not correct to suppose that each of the three prehistoric inventions discussed in this book was made by a single brilliant individual (analogous to Alexander Graham Bell). Rather, it is more likely, as we already have seen from the example of how dogs became the first non-human animals to be domesticated, that each of those inventions arose gradually over time, in a more or less accidental, unconscious, and non-intentional fashion, from a group of people who happened to share a certain attitude and way of looking at things. Nevertheless, it still is appropriate to speak of all three of those developments as inventions, because it probably is correct to say that each of them is an expression of various interests, values, ideas, and personalities of a distinctive set of people who lived in the past, who then passed down what they had discovered to other, later members of their societies.

4.3 A semi-digression: Talking does not have to be associated with counting

Counting is the religion of this generation; it is its hope and salvation.

Gertrude Stein

Homer's story of the *Odyssey* ends the following way. After a long and hard campaign of fighting in the Trojan War, and following that, an equally long, arduous, and dangerous return trip during which all his companions were killed, and he himself narrowly escaped death innumerable times, Odysseus finally succeeds in reasserting his authority, and reuniting himself with his wife Penelope, in their island home of Ithaca. Soon afterwards, however, he tells Penelope that he must undertake still another journey. This second journey, he says, is one he will take alone, on foot, and carrying an oar on his shoulder. It only will end when he comes to a place where the first people he meets will look at the oar and say, "What is that?" At that point, Odysseus says, he will bury the oar deep in the ground, and then turn again towards Ithaca, having found peace at last.

How, more precisely, is it appropriate to describe Odysseus' motivation for taking this strange, second journey? My hypothesis is that, in the light of everything he had experienced up to that point in his life, he might have been haunted by the depressing thought that all humans were fated to spend their lives in an endless struggle, both against one another, and against the elemental forces that were present in the natural world. However, if he could find even a few people who never had known the back-breaking, perilous work of using oars to push a boat through water, then that would provide grounds for suspecting that such a struggle was not inevitable after all. In other words, this discovery would support the idea that at least some people had retained the "natural," "unaltered," "innocent" state with which they were born, so that those individuals (unlike most others) still had their gods-given freedom to choose from among a wide range of styles of life. It also would show that humans, considered in general, did not have to live their lives in any specifically determined way, just because of the circumstances into which they happened to be born. To express the same thought in another fashion, my proposed interpretation of Homer's final story within his story is that Odysseus was searching for "natural noblemen" who had not allowed themselves to become enslaved either to accidents of history, or to the artificial constraints and demands imposed by the society in which they lived, because they had not lost contact with the full range of possibilities and powers that were contained in their originally given nature.

In a fashion roughly analogous to the preceding story, there was a flurry of excitement a few years ago, not just in scientific journals, but also in popular media like newspapers, topical magazines, and television programs, because of a series of Odysseus-like experiments the Columbia University psychologist, Peter Gordon, had conducted on members of a small Amazonian aboriginal, mainly hunter-gatherer tribe called the Pirahã. The reason those experiments were exciting is that they seemed to expand our conception of what is and—under some circumstances—is not possible for humans. For example, the experiments apparently confirmed the truth, at least in one instance, of the strong version of Whorfianism—i.e. the idea, suggested about

60 years ago by the amateur linguist Benjamin Lee Whorf, that any person whose language did not include a certain (kind of) word also would not be able to have the corresponding mental concept that word either presupposed, or to which it referred. (See Strauss, 2004, and Ingram, 2004.)[3]

The tribe just mentioned has a population of less than 200 people, living in small villages of 10 to 20 each, along the banks of the Maici River in the Lowland Amazonia region of Brazil. They speak a very simple language; they have no art of any kind; they do not recognize any distinct words for colors; they do not farm and have no domesticated animals; and they use a barter system of exchange rather than money. But the most important fact about them, as far as Gordon's theoretical and experimental purposes were concerned, was that their language had a system of counting that was based on the notions of "one, two, and many."

The idea of there being languages of that sort has been with us for a long time. For instance, the author of one of the newspaper articles to which I just referred (Ingram) said that, when he was a boy, he read in one of the footnotes in a *Reader's Digest* magazine whose function it was to offer tiny but interesting snippets of information, about a primitive tribe (living in a place whose name he no longer remembers) that had such a language. As the years went by, however, he discounted that story (presumably because he never heard any further details about the tribe) as nothing more than a fanciful and unsubstantiated myth. But that situation changed when, in the year 2004, he and other people became aware of Peter Gordon's research.

What changes have taken place in the world, between the 1950's, when both I and the author of the newspaper article were young, and the end of the first decade of the twenty-first century, which account for the difference just mentioned? The main change relevant to the topics I am discussing in this chapter is that, in the middle of the twentieth century, there still were many relatively forgotten corners of the world that were inaccessible, unknown, and dangerous for outsiders to visit. But now, in the twenty-first century just after the beginning of the new millennium, almost all of those places have become pacified, well explored, and integrated into larger, political and technological environments. This in turn has made it possible for

[3] By contrast, weak Whorfianism is the banal thesis that nearly every person would accept, that languages at least have some influence—of various different sorts—on the thinking of the people who speak them.

people from outside to visit those places, and even to conduct serious, detailed, and well-publicized scientific investigations there.

To illustrate the kind of factual information Gordon's work has made available to a wide audience, let me now describe some of his experiments. Gordon tells us that the Pirahã language has distinct words only for the first two numbers: "hói" (falling tone = "one") and "hoí" (rising tone = two"). All numbers larger than two are designated as "baagi" or "aibai" (= "many"). Over the course of a little more than two years, he and his associates first designed, then also administered to a number of the speakers of this language, a standard series of eight tests of progressively increasing difficulty. The goal of those investigations was to see whether the contrast between these subjects' behavior under well defined experimental conditions, and the behavior most other people who now live on Earth (who are not members of that tribe) would have displayed in approximately the same circumstances, could reveal some general facts about how the human mind was organized.

In an article published in the journal *Science*, Peter Gordon presented a figure or diagram that summarized the experimental results he had obtained from a typical series of tests administered to a group of seven Pirahã villagers (see Gordon, 2004, p.497). Since a picture often is worth a thousand words, I have reproduced that figure and its caption in my text—see below. This diagram shows that, of Gordon's series of eight experiments (A through H), the first four (A through D) are matching tests of increasing difficulty. All four of those tests were conducted on the surface of a table, where the experimenter sat on one side and the subject on the other. The goal for each of the subjects was to duplicate, or "make the same," an array of objects on his or her side of the table (separated down the middle by a wooden stick), with an "example array" which the experimenter arranged on his side. The first and apparently simplest of the tests (A) was a "1-1 Line Match" where the experimenter laid out different numbers (1 through 10) of AA batteries in an evenly spaced line, and where the subject's job was simply to reproduce that same arrangement and number of batteries on his side of the table. In the case of this first matching test, Gordon found (in a manner apparently consistent with strong Whorfianism) that all seven of the villagers were able to perform correctly for each of the numbers 1 and 2, for which their language had corresponding words. (In fact, they also managed to line up right arrays of three batteries as well, even though their language had no separate name for the number 3.) But they performed much more poorly for the cases of four and five, where the proportion of correct responses dropped to only about 70%; and in the case of seven, their correct responses dropped further still, to just above 50%.

The second test (B) was a "Cluster Line Match" where, instead of lining up batteries to correspond with a neat line of batteries located on the experimenter's side of the table, the subjects' job was to line up batteries in an array that matched the number of a small cluster (not a line) of ground nuts (not batteries) that appeared on the experimenter's side. Here again, although the subjects gave correct responses for the numbers one and two, their proportion of correct responses for all numbers higher than that dropped off rapidly. In fact, for the numbers nine and ten, none of the seven villagers was able to give any correct response at all. The third test (C) was an "Orthogonal Line Match" where the subjects' task was correctly to line up the corresponding number of batteries as in the experimenter's array, not in the same direction as the batteries in the experimenter's array, but instead in a line at right angles to the experimenter's line. Again, the subjects' responses were correct for the numbers 1 and 2, but were much worse for all the higher numbers, and dropped to zero for the number 9. The last of the four matching tasks (D) was an "Uneven Line Match" where the "target" batteries no longer were presented in a neat and evenly spaced array, but instead in a line with uneven gaps in it. As usual, the results here were correct for the numbers 1 and 2, but worse for higher numbers. Nevertheless, a "clumping" effect appeared here, where subjects sometimes improved their results by treating groups of batteries as if they were single items—for example, correctly reproducing an array of eight batteries, by considering two groups of four batteries, as if they were just two single elements.

The remaining four tests in Gordon's series (E through H) were more diverse. The first of these (E), which Gordon entitled a "Line Draw Copy" test, was one that any reader of this chapter would have found easy, but which all the Pirahã subjects had great difficulty negotiating. (In fact, it resulted in one of the worst performances of all.) The reason for this was that the test required subjects to reproduce, by drawing straight lines with a pen on a piece of paper, the same number of such lines that the experimenter already had drawn

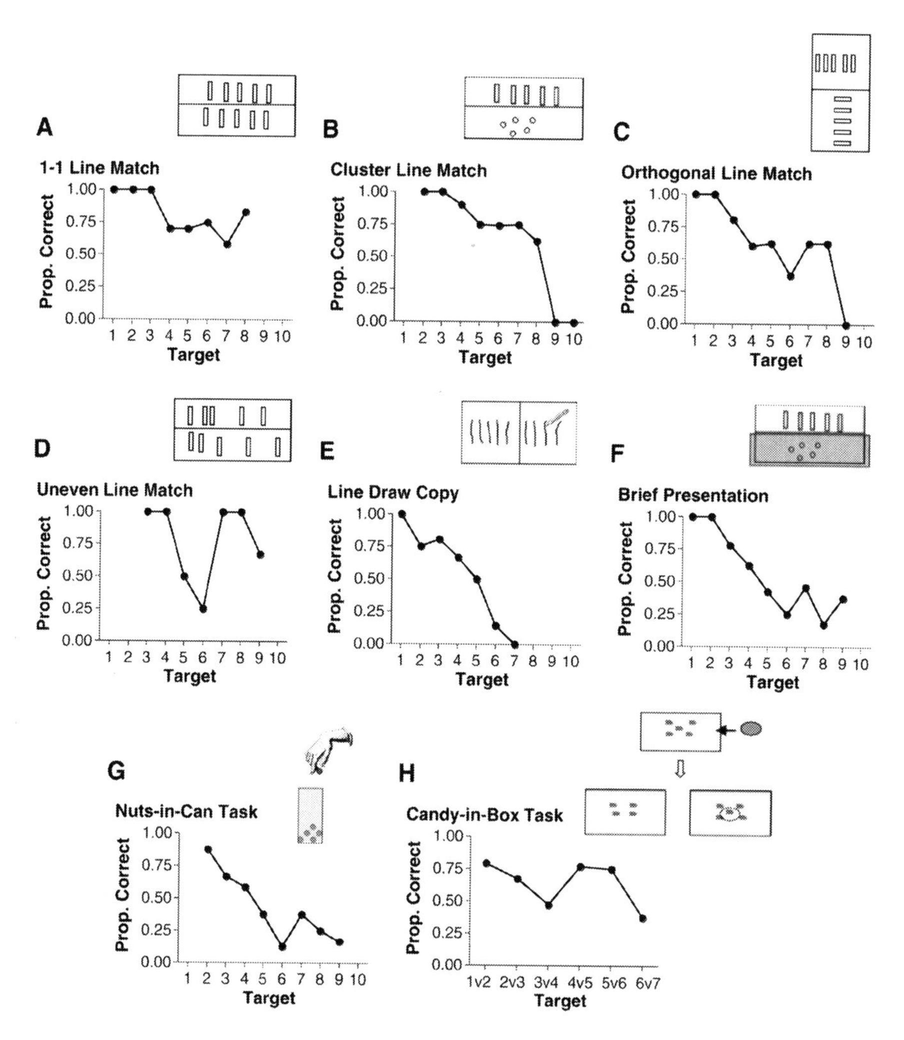

Figure 4.1 Results of number tasks with Pirahã villagers (n=7). Rectangles indicate AA batteries (5.0 cm by 1.4 cm), and circles indicate ground nuts. Center line indicates a stick between the author's example array (below the line) and the participant's attempt to "make it the same" (above the line). Tasks A through D required the participant to match the lower array presented by the author using a line of batteries; task E was similar, but involved the unfamiliar task of copying lines drawn on paper; task F was a matching task where the participant saw the numerical display for only about 1 s before it was hidden behind a screen; task G involved putting nuts into a can and withdrawing them one by one; (participants responded after each withdrawal as to whether they can still contained nuts or was empty; task H involved placing candy inside a box with a number of fish drawn on the lid (this was then hidden and brought out again with another box with one more or one less fish on the lid, and participants had to choose which box contained the candy).

on another piece of paper. But, as Gordon says (p.498), "Not only do the Pirahã not count, but they also do not draw." In other words, because all the members of our society have been trained to do things like this since childhood, none of us would have had the slightest trouble in doing it. But because the Pirahã had not been trained the same way, they did very poorly at it, and also experienced a great deal of distress—expressed in the form of "heavy sighs and groans"—in trying to accomplish the assigned task.

The sixth, "Brief Presentation" test (F) was the same as the second, "Cluster Line Match" test (B) already described, except for one thing. The one differentiating point was that, instead of allowing subjects to examine the cluster of ground nuts, they were asked to duplicate with a corresponding line of batteries, as closely, and for as long a time as they liked, they now were allowed only one second to see that array of nuts. That change—beginning with the set size three—had an immediate and dramatically negative effect on their experimental performance.

The final two tests (G and H) were both "out of sight" cases. That is, G, the "Nuts-in-Can Task," consisted in putting a bunch of nuts into a can, then taking them out one by one, and asking participants to say, after each withdrawal, whether they can still contained any nuts, or whether it was empty. In H, the "Candy-in-Box Task," there was a direct reward for good performance. Here the experimenter put candy inside a box with a certain number of fish drawn on the lid. The box was then hidden, and after that it was brought out again along with a second box that had either one more or one fewer fish drawn on its lid. At that point, subjects were given the task of guessing which box contained the candy. And, as the diagram shows, the results for these last two tasks were very bad—in a way, worse than for any of the others—since here subjects did not perform well even in the case of the numbers one and two and, for many of the higher numbers, they performed at a level that was no better than chance.

How did Gordon summarize the results of these studies, in his article? His main conclusion (see 2004, pp.498-9) was that his experimental tests had confirmed an instance of strong Whorfian linguistic determinism, because they showed that the Pirahã language was "incommensurate with" languages such as English, French, and Japanese, which allowed for exact counting and enumeration. He also claimed that the Pirahã subjects he tested showed evidence of using processes of analogue magnitude estimation, which allowed them to keep track of the different sizes of various sets of objects, by mentally comparing those sizes to one or another analogous, continuously varying magnitude, like the length of a line or the volume of a sphere. He speculated that the reason subjects did well in cases with a "target" group of one, two, or three objects was that the "analogue competence" of estimating the numbers 1, 2, and 3, did not rely on words at all. In his view, competence with those particular numbers was just a basic fact about human mentality. Gordon then posed the question of whether the Pirahã (who were different from most people on our planet today, because they had not learned, and had not been exposed to, either a number system or a numbering routine during the time they were growing up) could represent exact quantities for medium-sized sets of four or five objects. His tests showed that the answer to this last question was no.

These conclusions are similar to the ones that Pica, Lemer, Izard, and Dehaene also drew in an article that was published in the same issue of *Science* where Gordon's article appeared (see 2004). The four French researchers just mentioned conducted roughly similar investigations of members of the Mundurukú tribe of the Amazonian region of Brazil, whose language did not have number words beyond five. Like Gordon, their results showed that, although the Mundurukú were able to compare, add, and subtract large approximate numbers that were far beyond their naming range, they could not understand or make use of exact arithmetic for any numbers larger than 4 or 5. In a striking statement made towards the end of their paper (see 2004, p.503) these researchers said that all non-disabled Western children, at about the age of 3, exhibited an abrupt change in the way they processed and thought about numbers, because they suddenly came to realize that each of their count words referred to some precise quantity. But this same kind of "crystallization of discrete numbers out of an initially approximate continuum of numerical magnitudes" apparently never did occur among the children of the Mundurukú.

A point I now want to make is that the very existence of people like the Pirahã and Mundurukú constitutes a direct challenge to one of the principles of the Chomskean theory of language. Chomsky assumes that our grasp of arithmetic is directly based on our innately given ability to use and understand language. He says:

> We might think of the human number faculty as essentially an "abstraction" from human language, preserving the mechanisms of discrete infinity [i.e. infinity of the sort that belongs to natural numbers] and eliminating the other special features of language (1988, p.169).

Language is recursive, in the sense that speakers have a power to nest phrases within the scope of other, larger phrases, apparently without limit. (A simple example of this is as follows: "John believes that Mary believes that Ralph believes that Susan believes that Egburt believes that Nigel believes that Cassandra believes [and so on indefinitely] . . . that snow is white." See Pica *et al.*, 2004, p.499, and Hurford, 1987.) But to suppose as Chomsky does, that language, as opposed to an evolutionarily ancient ability to process numbers without the help of symbols, must be the source of our sense of recursion, begs the question. That is, this style of talking sets up the problem in a way that allows for only one possible solution, like the child who says, "Heads I win, tails you lose." Furthermore, there is at least one well-confirmed empirical fact that supports the anti-Chomskean notion that arithmetical ability (including the ability to count) is independent of language rather than derived from it. The fact of which I now am speaking is that, even though non-human animals have no language, many of them possess a number sense that is roughly similar to our own. For instance, although crows cannot speak in the same way as modern humans do, some animal behaviorists claim—as noted in the previous chapter—that they are more intelligent than non-human creatures of practically any other sort. One factor that leads observers to attribute a high level of intelligence to these birds is their ability to estimate the size of quantities (and in this respect, to count), at least up to the number six. (E.g. see Savage, 1995, p.4.)[4]

---

[4] Candace Savage says (1995, p.4):

> From the point of view of science, a raven is nothing more or less than an oversized crow. Jacob was an ordinary raven that lived in the laboratory of a German ethologist named Otto Koehler. For reasons of his own, Koehler believed that birds might have an innate sense of number. To test this conjecture, he designed a series of experiments in which Jacob the Raven was presented with a group of two, three, four, five or six small objects. Beside this "key" was placed a set of five small boxes, each identified with two, three, four, five or six black marks. The objects in the key and the marks on the boxes differed in size, shape and positioning and were changed from trial to trial. Jacob's task was to match the number of objects with the number of marks on one of the lids, open the correct box and obtain a food reward. In trial after trial, this is exactly what he did. Without the aid of language, Jacob the Raven could count to six.

Thus, Gordon and his associates, as well as the French team, have drawn attention to certain basic mental powers that belong to all present-day members of our species. For example, their investigations show that non-damaged homo sapiens have the first and, if conditions are right, also may have the second, of two types of numerical abilities. The first of these is the ability to estimate sizes and quantities in an approximate fashion. The second is the ability to refer to and work with exact numbers larger than three. The first power seems to be something that is innate in, and inherited by, every present-day human. Therefore it follows that only the second can be affected by either the presence or absence of number words in a speaker's language.

In spite of these researchers' useful work, I do not agree with the particular way they propose to interpret their data. More especially, I see no good reasons to believe that experimental observations like the ones summarized before provide support for the view known as strong Whorfianism. In my opinion, it was not just the Pirahãs' and Mundurukús' languages that caused those people to respond in the ways they did under the conditions of the experimental tests. Instead, what made them behave in these ways was something more general—namely, their entire societies, histories, and cultures.

Strong Whorfianism implies the literal truth of statements like one I once saw printed on a poster at the Toronto branch of the language school called *Alliance Francaise*—namely, "There are some things you only can say in French." In other words, this doctrine is that languages have such a powerfully determining effect on the thoughts of their speakers, that one cannot replace various words and phrases that occur in one language with (or "translate them into") corresponding words in another language. Since Gordon considers the language of the Pirahã to be "incommensurable" with every language that has an explicit counting system, he believes his investigations have shown that one clear instance of that principle is correct. But this conclusion only would be justified, if there were some natural, neat, self-consistent, and universally agreed upon means of dividing any given language (including all the things Gordon identifies with its "proper parts") on one side, from the cultural environment in which the language is set on the other. That is, Gordon shows that he has a

---

(I also have discussed the subject of counting in non-human animals, in my paper, "'Brutes Believe Not'," 1988.)

certain amount of sensitivity to the problem to which I now am referring, when he claims (2004, p.496) that terms like "molecule" and "quark," which would not exist in any culture that lacked advanced scientific institutions, cannot be considered to be proper parts of English, because people spoke English long before the time when those terms were invented. Accordingly, he says, failure to understand what a molecule or quark is, cannot be a sign that a person does not understand English, because English is only constituted by its "basic vocabulary and grammar."

However, the trouble with this way of talking is that languages constantly change; and therefore it follows, as Chomsky has reminded us, that none of them possesses an "essence" that all of its speakers accept, and upon which they all are able to agree. Consider a trivial example. In order to improve my shaky command of the French language, I sometimes listen to an audio-cassette of a French translation of Lewis Carroll's *Alice in Wonderland*, while driving my car. In that translation, whenever the size of Alice's body changes as a result of her having eaten one or another bit of magic mushroom, the new size she takes is consistently reported in terms of meters and centimeters rather than (as in the story's original, English version) in feet and inches. Were the makers of this cassette right to assume that the metric system of measurement is part of the "basic grammar" of the French language, and therefore one always must refer to meters and centimeters rather than feet and inches in any proper French translation? In my opinion, there are no clear and unambiguously correct answers either to that question, or to a very great number of other questions that are similar to it.

In particular, is it true to say that memorized counting procedures are essential parts of the basic grammar that belongs to certain languages? Or are procedures of that kind nothing more than extra-linguistic activities that some speakers have been fortunate enough to learn to employ, on strings of words or sounds that are contained in the languages they speak? Again, I see no reason to agree with Gordon's assumption that it is possible to draw a sharp line between language, on one hand, and various words, constructions, activities, behaviors and routines that are external to language on the other. Furthermore, even if we grant, for the sake of argument, that it is both feasible and legitimate to recognize some distinction of this sort, it is not clear to me that Gordon or anyone else is justified in locating words like "molecule," "quark," "feet," and "inches" outside the

English language, while putting memorized counting procedures inside it.

Thus in my opinion, theorists who maintain that languages are independent entities that do or at least can have a determining influence on culture are putting the cart before the horse. In my view, it is more realistic to conceive of each language as a constantly evolving expression of the particular cultural tradition with which it is associated. For example, it is appropriate for us to ask; (i) In what physical environment do the Pirahã live?; and (ii) what past, culturally influenced choices did they make, which led them to adopt the type of language they now speak? The answer to the first question is that they live in an undeveloped and undomesticated jungle and river environment, which has prevented them from becoming (or at any rate which was not conducive to their becoming) herders, or farmers, or traders, or soldiers, or money users, or writers, or artists, or historians, or scientists—any of which professions would have required them to develop skills of exact enumeration. Furthermore, long-term observations of the Pirahã by anthropologists like Daniel Everett (see the references to him in Strauss, 2004) make it clear that the Pirahã take a certain kind of pride in rejecting all the practices, professions, and pursuits just mentioned, in spite of the fact that they are constantly exposed to things of that sort from their interactions with the outsiders who surround them. Everett (quoted by Strauss) maintains that the Pirahã case demonstrates a fundamental cultural principle working itself out in language and behavior. The principle is that the Pirahã see themselves as intrinsically different from, and better than, all the people around them. And therefore, everything they do is a means by which they resist and try to prevent themselves from becoming like everyone else, and from being absorbed into the wider world. One way they have chosen to do this is by not making or forming abstract conceptions of anything: (e.g.) numbers, colors, or future events. This, according to Everett, is the basic reason the Pirahã have survived as Pirahã, while the great majority of the tribes around them now have been absorbed into the wider Brazilian culture.

Despite the culturally-based attitude about which Everett speaks, Gordon claims (on the basis of various statistical tests carried out on his data) that all of the Pirahã subjects he tested evidently wanted to perform well, and tried their best to do so, whenever they participated in the tests. Does this last point imply, in a manner dictated by Whorfian theory, that the language spoken by the adult members of the Pirahã tribe has made it impossible for them to think in the same ways as the ways that are characteristic of the readers of this book? I do not think the answer is yes. Consider, for instance, the "deep sighs and groans" that issued from the Pirahã subjects' mouths whenever they tried to draw straight lines on pieces of paper, in response to a request to duplicate the lines the experimenter already had drawn on another piece of paper. The symptomatic meaning of those groans, I suggest, is that they were signs of the fact that, at the time in question, those people were taking a first few steps down a long and painful path which, if they had been willing and motivated to follow it to the end, finally would have resulted in their learning to think in quite a different way than they do now. At least potentially, in other words, I claim, as against that which Whorf supposed, that the Pirahã implicitly have abilities to do all the same things as we do. But—of course—they could not learn to think in such a revised fashion immediately, or in just a single step, without their first having gone through a lengthy process of both training and retraining themselves.

What main conclusions have we reached in this section of the chapter? Neither Chomsky's reasoning, nor the experimental discoveries made by anthropologists like Gordon and the team of Pica *et al.*, establish the point that language is the foundation for everything else contained in present-day human nature. (This point remains true, even if it also is necessary to grant that language has important, and sometimes indispensable roles to play, relative to certain parts of that nature—e.g., to the ability to count up to numbers over a hundred.) As opposed to the viewpoint Chomsky has supported over his career, I do not think any theorist who proposes to arrive at a correct understanding of present-day human nature, should take language as his or her main concern. Instead, it is appropriate to pay attention to the cultures and cultural practices with which languages are associated, and out of which they arose, because the least misleading way to think about language is to understand it as an expression—admittedly, a central and important one—of a cultural tradition.

4.4 A key for distinguishing speech from codes (and thus also from the communication systems employed by many non-human animals) is to remember that the most important function of language is to enable subjects to think in new ways

"Two heads are better than one."

Chris Weyers, who says he talks to himself at work at a U.S. financial-service company

In a manner once favored by Darwin himself, many of his present-day followers—e.g. Jared Diamond (1992), Richard Leakey (1994), Steven Pinker (1994, 1997), and Philip Lieberman (1991, 2000)—subscribe to an evolutionary view of language that presupposes the truth of the following two principles: (i) there is no "difference of principle," which separates humans from organisms of other kinds, and—as one important instance of that point—(ii) it is impossible to draw a sharp line between language on one side and the systems of communication that quite a few non-human creatures use on the other. I do not believe that either of those presupposed claims is true. More especially, I intend to argue in this section that the second of them (and therefore also the first) must be false, if theorists like Chomsky are right to say that the most important job language performs is not to help two or more humans exchange information; instead, its primary job is to act as a means that allows each single, language-using person to engage in various forms of innovative and internal processes of thinking, just by himself or herself. To express the same point another way, if the general idea just mentioned is correct, it provides us with a means of distinguishing language on one side from codes on the other, on the grounds that, although language is an instrument that allows one to think in new and useful ways, codes are not. Furthermore, this same point also allows us to draw a relatively sharp distinction between (i) language, and (ii) the code-like systems of communication habitually used by many non-human animals, and by pre-UPR members of our own species.

(There are two slightly different forms of the Darwinian or evolutionary view of language, to which I am opposed. The first is based on the notion that the special factor that is essential to language occurs in many different versions and places throughout nature; and therefore many non-human animals possess a relatively simple form of language, which only differs in complexity from the language that belongs to mature and undamaged humans like the readers of this book. The second, roughly similar form of the Darwinian account of language says there is no one special thing, property, or type of things that is associated with the use of language; and therefore the human communication systems to which we arbitrarily give the name of "language" do not differ in significant respects from the means by which other creatures communicate.)[5]

An interesting historical example that provides support for the idea that observations do not confirm the truth of simple Darwinian assumptions about language, and thus about how human language originally came into existence, is the troubles that Japanese code-breakers had with a set of cryptic systems used by the United States Marines, in their "island hopping" campaign in the Pacific during the Second World War. The code-breakers to whom I now refer were faced with the puzzling fact that, no matter how hard and long they tried, their usual techniques for analyzing repeated patterns of sounds contained in passages of code they heard on the radio, did not allow them to decipher those passages. (In fact, some historians say this was the only code any Allied military organization used during the war, which the Japanese never succeeded in breaking.) Furthermore, the Japanese experts who worked on this so-called code even found it difficult to distinguish its individual sounds and phonemes, in such a way as to allow them to re-identify those sounds and collect them together again in repeatable sequences. For example, after the war, some of them reported that their impression at the time was that the American speakers they heard on the radio were "talking under water."[6]

To see what this means, let us consider some parts of the U.S. Navy's official website that bears the title, "Navajo Code Talkers: World War II Fact Sheet."[7] First, Navaho code talkers took part in

---

[5] Still another person who subscribes to either the first or second version of the Darwinian view of language is the linguist Steven Roger Fischer. According to his view, it is wrong and ultimately impossible to specify exactly when language came into existence, because (1999, p.8) ". . . any living being, in any epoch, that has used some means of conveying information to other animates has used 'language' of some sort. [Thus it follows that l]anguage is apparently a universal faculty."

[6] Where did I get this piece of information? I confess that it comes from a not very scholarly source (which I hope at least is basically accurate)—namely, the screenplay of a 2002 Hollywood movie staring Nicolas Cage, called "Windtalkers."

[7] My students, Jason Kennedy and Martin Veser, helped me locate this information.

every assault the U.S. Marines conducted in the Pacific between 1942 and 1945, including Guadalcanal, Tarawa, Peleliu, and Iwo Jima. For instance, Major Howard Conner, 5th Marine Division signal officer, said that if it had not been for the Navajos, the Marines never would have taken Iwo Jima. Conner had six Navajo code talkers working around the clock during the first two days of that battle; and in that time, those individuals sent and received over 800 messages, all without error.

I now shall quote several passages from the same website, on the grounds that they explain matters at least as well and clearly as my paraphrases could do:

> The idea to use Navajo for secure communications came from Philip Johnston, the son of a missionary to the Navajos and one of the few non-Navajos who spoke their language fluently. Johnston, reared on the Navajo reservation, was a World War I veteran who knew of the military's search for a code that would withstand all attempts to decipher it. He also knew that Native American languages—notably Choctaw—had been used in World War I to encode messages.
>
> Johnston believed Navajo answered the military requirement for an undecipherable code because Navajo is an unwritten language of extreme complexity. Its syntax and tonal qualities, not to mention dialects, make it unintelligible to anyone without extensive exposure and training. It has no alphabet or symbols, and is spoken only on the Navajo lands of the American Southwest. One estimate indicates that less than 30 non-Navajos, none of them Japanese, could understand the language at the outbreak of World War II.
>
> Early in 1942, Johnston met with Major General Clayton B. Vogel, the commanding general of Amphibious Corps, Pacific Fleet, and his staff to convince them of the Navajo language's value as code. Johnston staged tests under simulated combat conditions, demonstrating that Navajos could encode, transmit, and decode a three-line English message in 20 seconds. Machines of the time required 30 minutes to perform the same job. Convinced, Vogel recommended to the Commandant of the Marine Corps that the Marines recruit 200 Navajos.
>
> In May 1942, the first 29 Navajo recruits attended boot camp. Then, at Camp Pendleton, Oceanside, California, this first group created the Navajo code. They developed a dictionary and numerous words for military terms. The dictionary and all code words had to be memorized during training.
>
> Once a Navajo code talker completed his training, he was sent to a Marine unit deployed in the Pacific theater. The code talker's primary job was to talk, transmitting information on tactics and troop movements, orders and other vital battlefield communications over telephones and radios. They also acted as messengers, and performed general Marine duties.
>
> ....

The Japanese, who were skilled code breakers, remained baffled by the Navajo language. The Japanese chief of intelligence, Lieutenant General Seizo Arisue, said that while they were able to decipher the codes used by the U.S. Army and Army Air Corps, they never cracked the code used by the Marines. The Navajo code talkers even stymied a Navajo soldier taken prisoner at Bataan. (About 20 Navajos served in the U.S. Army in the Philippines.) The Navajo soldier, forced to listen to the jumbled words of talker transmissions, said to a code talker after the war, "I never figured out what you guys who got me into all that trouble were saying."

In 1942, there were about 50,000 Navajo tribe members. As of 1945, about 540 Navajos served as Marines. From 375 to 420 of those trained as code talkers; the rest served in other capacities.

...

The Navajo Code Talker's Dictionary

When a Navajo code talker received a message, what he heard was a string of seemingly unrelated Navajo words. The code talker first had to translate each Navajo word into its English equivalent. Then he used only the first letter of the English equivalent in spelling an English word. Thus, the Navajo words "wol-la-chee" (ant), "be-la-sana" (apple) and "tse-nill" (axe) all stood for the letter "a." One way to say the word "Navy" in Navajo code would be "tsah (needle) wol-la-chee (ant) an-keh-di -glini (victor) tsah-ah-dzoh (yucca)."

Most letters had more than one Navajo word representing them. Not all words had to be spelled out letter by letter. The developers of the original code assigned Navajo words to represent about 450 frequently used military terms that did not exist in the Navajo language. Several examples: "besh -lo" (iron fish) meant "submarine," "dah-he -tih-hi" (hummingbird) meant "fighter plane" and "debeh-li-zine" (black street) meant "squad."

The following passage is still another quotation relevant to the subject we now are discussing, which I have taken from the website—http://en.wikipedia.org/wiki/Code_talker:

An unfamiliar spoken human language is harder to crack than a code based on a familiar language. The languages chosen had no written literature, so researching them was impossible. Also, many grammatical structures in these languages are quite different from any the enemies would be expected to know, adding another layer of incomprehensibility. Non-speakers would find it extremely difficult to accurately distinguish unfamiliar sounds used in these languages. Additionally, a speaker who used the language all his life sounds distinctly different from a person who learned it in adulthood, thus reducing the chance of successful impostors sending false messages. Finally, the additional layer of an alphabet cipher was added to prevent interception by native speakers not trained as code talkers, in the event of their capture by the Japanese. A similar system employing Welsh was used by British forces, but not to any great extent.

What are my thoughts about this odd and, in some respects, nearly unprecedented case we are considering here? Part of the answer is that I consider it wrong, or at least misleading, for historians and other commentators to describe what the Marines were using as an "unbroken code," and for them to say things like "[a]n unfamiliar spoken human language is harder to crack than a code based on a familiar language." I say this because it strikes me as an obvious point, which every reflective person ought to accept, that a language—any language—is not a code at all. There are many facts that indicate the truth of this idea. For one thing, although a language can convey a potentially unlimited number of messages, none of those messages is hidden, in a sense that would require hearers familiar with the language to reconstruct that message by employing a series of indirect procedures. For example, the basic job Navajo code talkers were required to perform for the Marines was simply for them to speak with one another over the radio, in a manner they found comfortable and unconstrained, in their first-learned, "mother" tongue. Still more specifically, their duty was to engage in an unforced exchange of thinking about certain topics, in terms that were basically appropriate to their cultural background, and then to formulate and express certain conclusions arising from that exchange, in the words, accents, dialects, and sounds that were characteristic of their mother language and of the cultural tradition in which they had learned to participate as children. Of course, once the Japanese had taken a few Navajo prisoners whom they could force to help them get a grasp of the Navajo language and culture, it then became necessary for Marine instructors to introduce cipher-like elements into the conversations of the code-talkers, as an additional means of disguising what was being said. In spite of that, however, as compared to the Japanese experts' near mastery of English, which they had gained over many years of study, all their efforts to get a sufficient amount of familiarity with the Navajo language so as to comprehend what was going on in the broadcasts they heard, turned out to be "too little too late."

This case shows there is no "Darwinian continuum" that connects languages with codes. Instead, it is possible to draw a fairly sharp line between items of these two sorts, in the view of the fact that languages—unlike codes—are not explicitly planned systems of artificially stipulated rules. For instance, we know from ordinary experience that people learn and employ languages in a very different fashion than they decipher codes. The fundamental thing a person does to acquire a language is to go through a long and repetitive process of "getting used to it." For example, any adult who reads the bedtime story to children, about Chicken-Licken, Goosey-Loosey, Ducky-Lucky, and Turkey-Lurkey all going to tell the king that the sky was falling, soon becomes aware of the fact that children need a great deal of repetition to become competent users of English. Furthermore, if that same adult then enters a program of learning some previously unknown language himself or herself, he soon discovers that he also needs exactly the same kind and amount of repetition as children do, in order to accomplish this goal.

What is it that happens when a person familiarizes himself with a language? What kind of change does this process bring about? The answer seems to be that repetitive practice with the elements and forms of a language gradually transforms the brain of each individual learner, so that the learner finally comes to be in a state that allows him or her to understand and respond appropriately to the sounds of the new language in an automatic fashion, without having to think about any technical details such as grammatical rules or vocabulary.[8] Thus, people—including Japanese linguists listening to strings of Navajo words on the radio—who have not previously engaged in a long process of sharpening their skills of comprehending, responding to, and speaking Navajo by means of practice, will find that they are not even able to hear and distinguish the various sounds of that language in a clear, reliable, and accurate manner.

To be being even more explicit, let us distinguish between (1) a typical way of thinking and expressing some thought **E**, which is characteristic of speakers of English; (2) a typical way of thinking and expressing a thought **J**, which is characteristic of speakers of Japanese; and (3) a typical way of thinking and expressing a thought **N**, which is characteristic of speakers of Navaho. The Japanese code-breakers' method was to look for **E**-like patterns of thinking, expressed in the sounds they heard from the mouths of the American code users, and then to use those patterns as a basis for reconstructing the disguised English messages they believed must underlie the coded passages in

[8] In Lewis Carroll's *Alice's Adventures in Wonderland*, the Duchess makes the wise remark to Alice (which she afterwards presents to her as a gift), "Take care of the sense, and the sounds will take care of themselves."

question. But this method was bound to fail because (at least for the most part), when they were confronted with the sounds of the Navaho language being spoken by native speakers, there were no genuine **E**-like patterns of thought present in what the Americans said, but only **N** patterns which, of course, the Japanese had no way of recognizing. The reason for that in turn was that the Japanese were not hearing a code at all, but instead were simply hearing native speakers, speaking in their own language.

In the case of the vast majority of non-human animals, I do not believe the sounds they make constitute a language of any sort, not even a very simple one. The reason I believe this is that those sounds are not a means by which the animals express thoughts and make choices, but are merely ways in which they have been determined to behave by ordinary natural selection. In my view, a language is something that comes into existence as a result of a certain kind of cultural and artificial selection, in contrast with and as opposed to natural selection, because language is a means that speakers use to express their own intentions. By contrast, sounds like bird-calls, which are determined by natural selection, are—as the phrase "natural selection" implies—items that nature has "chosen" for, or on behalf of, the animals that make those sounds.

Let me summarize the preceding point as follows. The crucial thing about a language is that it functions in something like the same way as an additional part, aspect, or member of the person who learns it (like a new arm). But the same is not true of a deciphered code. For example, whenever someone speaks in a spontaneous, unconstrained, and honest fashion (as opposed—for example—to his or her operating as, or as like, a real estate agent trying to sell a house, or as an actor on a stage, or as an applicant for a job, or as a student taking a *viva voce* examination, or as a person who is being threatened with torture, etc.), hearers are entitled to interpret what that speaker says as an expression of his own ideas, values, and personality. This is because the words they hear him or her speaking issue from thoughts he either has now or has had in the past; and those same thoughts also are based on choices he either has made or is now in the process of making. By contrast, codes are and always remain "external" to their users, since their purpose is not to reveal what someone considers to be the "real meaning" of the objects, facts, and situations with which he or she is concerned, but instead is to disguise and bury such meanings, by making it difficult—even comparatively difficult for those who possess the code's key—to recover them. For instance, the military codes invented by Julius Caesar (one of the first people to employ such things) were simple systems for substituting and reshuffling letters, which were meant to accomplish the goal of scrambling, and thereby concealing, messages written in ordinary Latin. Thus, what we now refer to as a "Caesarian cipher" is a means of reordering the 26 letters of the Roman alphabet, by putting a previously agreed code word at the beginning of the alphabet, and then continuing on with the rest of the letters in their usual order, minus the letters that occur in the code word. Here is an elementary example. Suppose the code word chosen for the cipher—which is known by both its senders and its receivers—is "rex" (the Latin word for "king"). In that case, a sender would replace all instances of the letter "a" in his original written message with the letter "r," replace the letter "b" with the letter "e," replace "c" with the letter "x," "d" with "a," "e" with "b," "f" with "c," "g" with "d," "h" with "f," "i" with "g," "j" with "h," "k" with "g," "l" with "h," and so on.[9]

Let us now turn to another topic related to the matters we just have been discussing. What is it appropriate to say about the vocalizations and other sounds made by non-human animals, which those animals apparently use in at least a quasi-communicative fashion—e.g., the songs of birds, whales, and fish; the roaring of lions; the howling of wolves and howler monkeys; the cries of seagulls; the caws and croaks of crows; the sound a rattlesnake makes with its tail; the chirping of crickets? Are followers of Darwin right to say that all those behaviors are at least analogous with, and may even be simple approximations to, the speaking of a language? Or are there better and more accurate ways of describing them? I believe the second answer is correct, because none of the behaviors just mentioned is even remotely similar to a language. What I mean by saying this is that (a) none of those behaviors is an expression of thought, (b) none is based on choices, and therefore (c) all of them are "externally related" to the animals from which they issue, in the sense of not expressing anything distinctive or personal about those animals. In

[9] People who know more about this subject than I do have informed me that the era of deciphering military codes is now part of past history. This is because each of the vast majority of codes military people employ today is generated with the help of powerful computers that employ one or more randomizers. Furthermore, any such code always is replaced by another one, before the time it would take equally intelligent enemy experts, working with equally powerful computational tools, to "break" it.

fact, in view of the preceding points, I claim that most animal vocalizations are more like codes than they are like languages.[10]

Consider a concrete case. The English poet Robert Browning famously wrote:

> That's the wise thrush; he sings each song twice over,
> Lest you should think he never could recapture
> The first fine careless rapture!

The charm of these lines lies in the fact that they are a means of pretending that something is the case, which both the poet and his readers know is actually false. That is, the thrush does not repeat its song in order to affirm or prove something to its human hearers. Instead, all of us are implicitly aware of the fact that the real reason a thrush repeats its song is that it cannot stop itself from doing so.

Analogous to what I just said about codes, I suggest that there is an important difference between the way post-UPR humans like us acquire languages, and the manner in which members of other species produce (as well as the way earlier members of our own species once produced) the sounds that are and were typical of them. After going through a period of encountering the sounds of one or several languages being spoken around him or her, a young human child begins to believe in the "rightness" of certain interpretations he observes native speakers giving to those sounds. To say the same thing another way, he starts to think of the sound combinations just mentioned as words. Even more generally, the child accepts the idea that the special way in which his mother tongue(s)[11] designate(s) the entities in his environment is something that is appropriate, fit, and natural, since that way is an accurate reflection of "what things really

---

[10] I limit this claim only to "most" animals because it seems to me that researchers have proved (by means of various sorts of "double–blind" tests) that a few, highly trained, non-human animals have become able to understand and make use of something like meaningful, human-like language. For example, Sue Savage-Rumbaugh has shown this for the case of the bonobo named Kanzi (see Savage-Rumbaugh, Shanker, Taylor, 1998, pp.3-74) and Irene Pepperberg has demonstrated it for the case of the African Grey Parrot named Alex (see Pepperberg, 2008, passim, especially pp.103-8).

[11] I use the plural form here because children often learn two or more languages when they are growing up. For example, one of the early prime ministers of Canada reported that, as a child, he was not aware that there was one language called "English" and another language called "French." All he knew was that there was one way in which he had to talk to his mother, and another way in which it was necessary to speak to his father.

are." Furthermore, he also begins to define himself and his own personality—e.g. attitudes, prejudices, hopes, fears, and values—in ways that are consistent with the character and peculiarities of the particular language(s) and culture(s) that he has learned. By contrast, the vast majority of nonhuman animals do not go through any similar process of accepting and assimilating into their own personalities, the special characteristics of the sound patterns they hear, and of the ways they observe those sounds being used by other members of their species.

To return to a previous example, thrushes do not "recapture" their songs by first learning them and then accurately reconstructing them at a later time. Instead, even before those birds are hatched, they carry an innate "template" that eventually enables them to produce all the songs they will sing during their lifetimes. Admittedly, members of various species of thrushes, who live in different areas, often sing in recognizably different dialects. But in spite of that, the basic pattern for each of their songs (for example, that which one expert[12] has described as the *whee-wheeoo-titi-whee* call of the Grey-Cheeked thrush) is the same for all the members of that species. Thus, it is not true that heard sounds gradually acquire a meaning for thrushes. It also is not true that meanings that are associated with those sounds influence, and then finally become a part of, each of the personalities of the individual thrushes themselves. Furthermore (excluding cases like Kanzi and Alex), even the non-human animal sounds that strike us as being comparatively more human-like and language-like than most other cases—e.g., vervet monkeys' use of three different "words" to warn of attacks by leopards, snakes, and eagles—are not different in principle from the situation typical of thrushes.

As already noted in our discussion of linguistic "traces" in section 4.2, post-UPR humans like us have at least some thrush-like characteristics. That is, we also have innately given templates that dictate many of the properties that belong to all the languages that we are able to learn. For instance, one of those templates requires that every word and phrase that occurs in a properly constructed sentence of a learnable language should be organized hierarchically, so as to allow linguists to describe the structure of that phrase by means of a "tree diagram." Nevertheless, contrary to the views of "innatists" like Chomsky and Pinker, I do not believe—a point based on reflections on certain events scientists discovered to have occurred during our

[12] F.H. Allen. See Peterson, 1964, p.173.

species' pre-history—that those language-templates are sufficient, simply by themselves, to produce language (as opposed merely to supporting, undergirding, or helping to organize language), without the assistance of, or any important contribution from, culture.

Accordingly, then, the only kind of meaning it is sensible to associate with a code, or a code-like animal vocalization, is one that completely depends on—or to use another word for the same thing, that is parasitical on—various objects, facts, and situations that exist outside the code or vocalization itself. In a sense, therefore, it follows that all codes, and code-like animal calls and cries, are essentially similar to and interchangeable with one another. On the other hand, languages are far more varied, and are far more detached from the things with which they deal, than is true in the case of both codes and animal vocalizations. The reason for this is that each language (or to state the point more correctly, each person who speaks that language, and therefore who also thinks in and by means of it) interprets and describes its referents in terms of forms and conventions that are intrinsic to the language itself, and in ways that are consistent with the special attitude, style, and viewpoint that also are characteristic of that language.

Chomsky and the people who follow his leadership do not accept this idea, because they believe that all the languages that exist on Earth are basically similar to each other, by virtue of the fact that all their properties are dictated by biological and genetic processes that belong to our shared human nature. For example, Chomsky says (see1997a, p.25) that an extraterrestrial scientist who visited our planet would conclude that, except for a few trivial surface details, all human beings spoke the very same language. But this way of talking leaves out an important point. As opposed to Chomsky's view, it seems to me that every particular language has been influenced in significant and non-trivial respects by the special culture and history of the humans who speak that language; and because of that, each language counts as a significant and typical part of the human culture with which it is associated.

Let me give a simple illustration of this point. Consider what some military historians have said about the roles respectively played by the English and Japanese languages, in determining the outcomes of various battles that took place during the Second World War. Winston Churchill provided a quick summary of this view, by means of the following sentence:

> The rigidity of Japanese planning and the tendency to abandon the object when their plans did not go according to schedule is thought to have been largely due to the cumbersome and imprecise nature of their language, which rendered it extremely difficult to improvise by means of signaled communications. (1950, p.253.)[13]

Although I do not speak Japanese, my impression—formed from various facts present in the historical record—is that this statement contains a certain amount of truth. Nevertheless, I believe Churchill's sentence is far too simple. Which came first, the rigidity of the planning done by certain members of the Japanese army and navy in World War II, or the cumbersome and imprecise nature of their language? The right answer, I suggest, is that neither of these things deserves to be thought of as the cause of the other, or as having preceded the other, because both of the items just mentioned—not just in the past, but also continuing on into the present and future—are aspects of the same historical and cultural tradition. In other words, it is not just the Japanese language, but also its speakers, their history, and many of their culturally determined ways of thinking and behaving, which one should describe as (e.g.) precise, responsible, detailed, and thus also often rigid. (A further comment: What makes it sensible to say things of this sort about our fellow humans, but not about animals of other species? The simple answer is that the difference to which I am referring stems from the fact that human cultures are much stronger, deeper, more complex, and also that they are far more important determiners of behavior, than the cultures that are associated with other known creatures. On this point see e.g., Whiten and Boesch, 2001.)

All the codes about which we have been speaking up to this point in the chapter are "mixed" cases that only non-infantile, modern, literate humans can employ, because the target for referential substitution in those codes is pre-existing linguistic texts (messages). To be more specific, we have been speaking so far about military codes and ciphers which, as mentioned before, are systems designed to encrypt—and thereby to hide—phrases, formulas, and sentences contained in previously written messages. Nevertheless, it also is clear that some code-like ways of communicating do not make any use of language at all, since they are composed of what one might call

---

[13] My friend, David Weind, pointed out this sentence to me.

"natural signs."[14] Consider, for example, the bodily signals someone uses to prompt, lead, and guide either a dance partner or a horse. A rider (man or woman) can communicate effectively with a non-speaking horse, in roughly the same way that a male dance partner communicates with his female partner, because in both of these cases, the code used to do this is expressed in terms of bodily movements aimed at eliciting an appropriate, "answering" set of bodily responses.[15]

By the way, why does horsemanship remain a regular part of the training of cadet officers in military academies around the world today—long after horses have ceased to be our principal means of transportation? I believe the main reason for this is that the individuals whose job it is to organize the curricula of those educational institutions are aware of the fact, confirmed over thousands of years, that the mental and physical habits cadets must develop in order to ride horses, are also useful to them for the performance of their more important job of commanding troops and subordinate officers. Since horses cannot talk, all the techniques riders must master in order to acquire a satisfactory level of horsemanship are non-linguistic ones that only involve factors like bearing, poise, assurance, determination, authority, and other aspects of "body-language." (Of course, human body language often includes use of the voice as well. But the vocal effects to which we now refer might be accomplished just as well by gibberish, as by grammatically correct words and phrases contained in recognized languages like English, German, Russian, or Mandarin.)

In view of these "larger" implications of horsemanship, the most insulting thing Thomas Paine could think to say about George Washington—a person he detested—was that even though Washington often chose to address his soldiers from the back of a horse, the horse on which he sat was, according to Paine, only barely under his control. To be still more explicit, Paine's remark implied (and was meant to imply) that, if Washington was not even able to gain the respectful subservience of his own horse, then it could only be an

---

[14] Derek Bickerton claims that the various different types of what he calls "protolanguage"—e.g. pidgins, the speech of very young children, and the linguistic accomplishments of trained apes—are more or less like this as well. See his 1990, especially Chapter 7.

[15] I hope lady dancers with whom I am acquainted will not be insulted by the fact that I have mentioned these two cases in the same breath.

absurd fantasy to imagine that he also was qualified to serve as the first president of the new nation he supposedly represented.[16]

I have claimed that code-like systems that have nothing essential to do with language are the means non-human animals use (and which pre-UPR humans also once used) in order to exchange messages and, in this sense, to communicate. For example, certain movements in the dance of honeybees act as elements in the non-linguistic code those particular animals employ, which allows them to indicate (by means of simple substitutions) the direction of the sun from the hive, and the direction from the hive of a newly discovered stand of pollinating flowers. Again, it also is plausible to suppose that the patterns of sounds in the songs that are associated with male humpback whales (only the males sing) have roughly the same function as the mating display of brightly colored glass, feathers, leaves, etc. that are created by male bowerbirds in New Guinea. That is, the basic purpose of both those displays is to impress females with the intelligence, industry, esthetic taste, and other qualities of the males who created them. In turn, those same perceived qualities also are meant to be understood—and accepted—by the female humpbacks and bowerbirds, as signs that those particular males are desirable breeding partners. In other words, the males use these sounds and displays as code-like references to (i.e. substitutions for) various sexually desirable qualities they possess.

A problem we now face is this: How, at a certain time in the past, did humans succeed in moving away from their previously exclusive reliance on non-linguistic codes, and into a new situation where they became able to use language as well as codes? My general answer to this question is that language developed out of a special type of cultural tradition that only was available to modern humans—namely, a culture of creative innovation. To be still more specific, I suggest that language was invented, and thus came into existence, in roughly the same fashion as every other major human invention. For example, language was similar in that respect to controlled fire, cooked food, spear throwers, fishnets, needles, shoes, shoelaces, hobbles for horses, and the wheel. Furthermore it was an invention,

[16] Still another parenthetical remark: If bodily and verbal (gibberish) techniques work just as well for commanding language-using humans as for controlling horses, then it apparently must be the case that at least the broad foundation of human nature (as opposed to its more specialized aspects that distinguish and separate us from other living things) must be something we share with a large number of non-human creatures as well—including horses.

like all those just mentioned, which—given certain preparations—was both (a) highly useful and (b) relatively obvious. Because of this last point, we should not be surprised to learn from archeologists' reports that several groups of people succeeded in thinking of it independently, at different places and times.

Even though I conceive of language as a comparatively intuitive and natural invention, this does not imply (nor do I claim, in the style of theorists like Chomsky, Donald, and Mithen) that the development of language was inevitable, in the sense of being something that was bound to happen sooner or later. Rather I think that, in every separate case, language always began from the creative thinking that was done by some unusual and remarkable person, or a fairly small number of remarkable people, who also happened to be born into an equally remarkable set of circumstances. Of course, part of what it means to say that the people in question were remarkable, is that they were individuals who very well might not have existed.

In my view, then, codes are systematic means of substituting one set of things for another, which sometimes can be useful for communication of a certain sort, because of the fact that correlation and substitution often help us identify, refer to, and keep track of certain items in terms of others. Furthermore, it is appropriate to distinguish between codes of two general types: (1) those that are unlearned, like the cries of birds, the songs of whales, the dances of bees, human and animal body languages, etc., and (2) those that are invented and learned, like military ciphers, ballroom dancing, riding horses, and (perhaps) artistic expression. By contrast, it seems to me that Chomsky is right to say that languages are only incidentally useful for communication, since their primary function is to allow humans to think in certain ways. For example, their operation does not depend on systematic substitutions for things, persons, and situations. Instead, they are means that allow people to describe things, to explain them, and to make personal comments about them. Although codes often are useful, they are externally related to us. In this respect, they are similar to physical tools like hammers and pliers which, no matter how familiar and precious they might become, do not really count as parts of us. By contrast, since languages express and grow out of cultural traditions, they gradually become assimilated into a person's being and personality. In fact, as suggested before, learning a language is like growing an extra part of one's body, like another arm, leg, tongue, or—more especially—an additional (part of the) brain.[17]

4.5 Was Helen Keller right to believe she suddenly had been transformed from an animal into a human?

Helen Keller took a great deal of care to recall accurately, and to describe in detail in her autobiographical writings, a certain event she thought of as the most important thing that ever had happened to her. For example, she did not talk about what had taken place on a certain special day, hour, minute, and second in just one of her books, but in several. Furthermore, she repeatedly claimed that this event had changed her from being a mere creature into a fully conscious and functioning human being. For example, consider the following description she gave of the incident (1903/1996, p.12):

> We walked down the path to the well-house, attracted by the fragrance of the honeysuckle with which it was covered. Some one was drawing water and my teacher [Miss Anne Mansfield Sullivan] placed my hand under the spout. As the cool stream gushed over one hand she spelled into the other the word *water*, first slowly, then rapidly. I stood still, my whole attention fixed upon the motions of her fingers. Suddenly I felt a misty consciousness as of something forgotten—a thrill of returning thought; and somehow the mystery of language was revealed to me. I knew then that "w-a-t-e-r" meant the wonderful cool something that was flowing over my hand. That living word awakened my soul, gave it light, hope, joy, set it free! There were barriers still, it is true, but barriers that could in time be swept away. I left the well-house eager to learn. Everything had a name, and each name gave birth to a new thought.

When Helen was a small child, she had learned to speak the word "water," as part of her verbal communication with people around her, during the brief period before she caught the illness (possibly scarlet fever) that was to leave her totally blind and totally deaf. This presumably explains why she spoke in the passage just quoted about having a misty consciousness of something forgotten—a thrill of returning thought. Helen also said that as soon as she had acquired the meaning of the hand-signed word for water, she ran excitedly to

---

[17] Some physiologists say that learning a second language can be like taking out an insurance policy. This is because if, under certain circumstances, a person suffers a stroke that robs him or her of the use of his first language, the second one still might survive more or less intact.

one after another of many nearby objects, and demanded that Anne Sullivan teach her the words for all those other things as well. Thus, this single, meaningful word seemed to "unlock" things in a way that allow it to be joined by a potentially infinite number of other words of the same sort as well.

What did she mean when she claimed she had not really been a human before the incident at the well-house? She said in her book 1904/10, pp.113-6 (also quoted by Dennett in 1991, p.227):

> Before my teacher came to me, I did not know that I am. I lived in a world that was a no-world. I cannot hope to describe adequately that unconscious, yet conscious time of nothingness. . . . Since I had no power of thought, I did not compare one mental state with another.

However, if one supposes it was literally true that she was nothing more than an animal before that incident took place, then she must have been an unusually intelligent and resourceful animal at that time. This is shown by the fact that—before she had access to her teacher—she spontaneously had worked out a large system of behaviors and felt signs for the purpose of communicating simple ideas to people around her. For instance, a shake of her head (on which she had pressed the hands of the "hearer") meant "No," a nod meant "Yes," a pull on a nearby person meant "Come," and a shove meant "Go." If she wanted bread to eat, she would act out the actions of cutting the bread and then spreading butter on it with a knife. (See 1903, pp.4-5.) I never have heard of any non-human animal that independently had developed a system of body signs that was as large, subtle, imaginative, and human-like as that one.

Before the well-house incident, as a result of training she had received from her teacher, Helen already had learned about 30 tactile language-patterns that referred to various individual things. Among those patterns was the spelled word *baby,* which represented her younger sister, and *doll* that referred to a particular doll Anne Sullivan had given her. On the morning of the day of the incident, she had the new experience of learning from her teacher the tactile pattern *water*, which (as Helen then understood it) referred to the warm water she felt in her bathtub at the time when she was being scrubbed and groomed to get ready for the day. A few minutes later on that same morning, Anne Sullivan tried to get Helen to use the previously learned tactile pattern *doll* to refer to a new, second doll. But Helen so strongly and violently rejected the idea of using the same tactile pattern to stand for two different things, that she threw the unfamiliar doll onto the floor hard enough to break it. It was immediately after that—and also partly perhaps, because her tantrum with the doll had created a "teachable moment" in Helen's mind—that Anne Sullivan took her to the well-house, where the cold water on her hand led her to have a new realization.

What do I think was the fundamental point that was implicit in the "mystery of language" that Helen Keller claimed to have learned that day? I believe the crucial thing was that she came to understand that the meanings of words did not have to be limited, in a mechanical and code-like fashion, to the single job of standing for certain concrete and particular things. For example, "water" could refer both to the warm and calm liquid she felt in her bathtub and also to the cold liquid that came forcefully splashing out of the pump. In fact, she apparently grasped the point that the meanings of the vast majority of words—as well as all the words that were most central to, and typical of, language—were abstract and general, because language was not just a set of quasi-mechanical, 1-1 substitutions that a code-breaker might "crack." Instead, each word was a flexible instrument of thinking whose precise application and meaning in any situation had to be freely chosen by the language user. The reason this was true was that there was a very large number of possible ways in which a speaker could express a thought (in fact, there was a 1-infinity relation here); and speakers could appeal to any of those ways in order to formulate, clarify, expand, or merely to celebrate any idea that occurred to him or her, before he went on to the practical business of sharing that idea with other people. Still another way of summarizing this same point is to say that what Helen learned was that instead of words always having to be narrowly focused on corresponding objects, it was more accurate—and far more important—to understand that speakers (signers) were able to use words in a non-particularized fashion that allowed them and their hearers (recipients) to communicate about whole sets and combinations of relevantly similar items, even including things, events, and situations that were merely analogous to what one actually experienced (e.g. "the healing water of forgiveness"), or that were speculative and merely possible rather than confirmed as actual (e.g. "the water future space travelers hope to find in polar regions of the moon").

4.6 Our ancestors may have learned their first expandable word—and thereby also acquired their first full language—by means of a shared memory that became fixed in their minds through something like a divine revelation

> Whether we are talking about the origins of life or the origins of language in the hominization of primates, we are talking about a big mystery. All of these new ideas and new questions bring us to the edge of knowledge and a horizon in which perception and imagination, like earth and sky, meet.
>
> W. Thompson

As already noted, many observers believe the insightfulness and problem-solving skills of the birds called corvids—crows, ravens, magpies, jays, etc.—show that those birds are at least equal in intelligence, and in some respects are superior, to non-human primates like chimpanzees. In particular, the unusually wide range of sounds ravens make might be an indication that we are justified in thinking of that species as the most intellectually gifted of all the corvids. Candice Savage says (2005, pp.88-93):

> To date, the researchers have compiled a library of more than 64,000 vocalizations, elicited from 37 raven pairs. From this cacophony, they have distinguished 84 distinctly different calls, and the list continues to grow as each new pair of ravens is added to the choir. Within the limits of their syringeal organs, the birds appear to be free to learn, imitate, and invent, and their collective vocal repertoire is thought to be open-ended. Yet with all these possibilities at their disposal, each individual adult raven has a limited vocabulary of only about a dozen calls. . . . The only possible conclusion [is] that, with the probable exception of a few basic vocalizations like the begging cry of nestlings and the yell of feeding mobs, ravens' calls do not come with genetically preassigned functions. Instead, it seems that these clangorous utterances acquire their definition and meaning in the context of each raven's social experience. Although "ravenese" may not be a symbolic language as we know it, neither is it a simple system of signals, a collection of bestial grunts and groans. We're beginning to catch a glimpse of something beautiful.

I propose that, before the advent of full human language, our ancestors' vocalizations might have been something like the sounds made at the present time by ravens. Many calls in the repertoire of ravens are inherited products of instinct, and thus are more or less standardized ones that all the undamaged individuals in the species share (like humans' sound of crying and laughing). However, since inventiveness is as much a mark of intelligence as diversity, it also is plausible to suppose (following Savage) that at least some of the sounds ravens make are not shared ones. That is, certain calls of ravens are inventions that certain individual birds may have made, which therefore are expressions of certain thoughts and desires that are characteristic of some particular ravens, but are not common to all the members of that species. Presumably, those invented sounds are only employed (as well as revised, mixed, scrambled, and run together) by small family-based groups of birds associated with the particular inventor or inventors of those sounds, and therefore cease to be made, once all the members of that group have died.

In other respects, it seems to me that our pre-linguistic ancestors were more like wolves than like ravens. As noted in the preceding chapter, crows and ravens congregate in large flocks at certain times of the year—especially in winter. But even when they gather in this fashion, these birds continue to act in independent and individualistic ways, since the only organizing principle that appears to apply to the members of those flocks is what one might describe as "intelligence without leadership." But this is not what humans are like. Relative to this last mentioned characteristic of leadership, most humans act in ways that are more similar to the behavior typical of wolves than the behavior of ravens, since both human and wolf societies tend to take the form of relatively small, hierarchically organized sets or packs headed by either one or a small number of alpha individuals, which do not become submerged or dissipated if and when the members of that group come together in larger concentrations.

I furthermore speculate that pre-linguistic members of our human species probably made at least some non-inherited sound innovations, like the ones Candace Savage attributes to ravens, which were ephemeral in the sense of not lasting for a very long time. Thus, consider the following question: What finally allowed some humans to move beyond this raven-like situation, by inventing a language that, even though it was invented rather than brought into existence simply by genes, also could be passed down over a much longer span of time, from one generation to another?

Somewhat analogous to planets that exist in a double star system, modern members of our species tend to think and behave, and thus also to organize the groups in which they live, in a bi-polar fashion, because they recognize two different types of authority. Those two types of authority are respectively associated with chiefs, generals, and bosses on one side, and with priests, commissars, and seers on the other. The principal job and responsibility of a chief, which approximately corresponds to the role of the alpha wolf in a wolf pack, is to direct and control the relatively superficial events that make up what it might be appropriate to call "the foreground of life"—e.g. where group members will live, what and when they will eat, how they will defend themselves against rivals and other enemies, how their social hierarchy will be constituted, and so on. On the other hand, the authority of a priest (to which nothing corresponds in a wolf pack) is basically moral, spiritual, and interpretive rather than physical in nature. A priest's main job is to explain the past and present, predict the future, and try to bring the other tribe members into correct and wholesome contact with various mysterious powers that are assumed to control "life's general background."

Something else that seems relevant to the explanation of language is the claim Noam Chomsky makes that every genuine language should be characterized in terms of its relation (of one or another form) with "discrete infinity." Chomsky would deny that ravens and wolves are genuine language-users, on the grounds that although ravens make as many as sixty-four recognizably different sounds, and wolves employ a smaller but still relatively large number of significant sounds and gestures, neither of those species can produce a potentially infinite number of such signs. In terms of the example of counting discussed in section 4.3, although we know from observations that ravens and crows can remember and keep track of (i.e. count) as many as six to twelve different items, that is as far as their counting can go. On the other hand, there is no theoretical limit to the number of separate items language-using humans are potentially capable of recognizing. (This statement even applies to members of the Pirahã and Mundurukú tribes of Brazil, if one takes account of approximate as well as definite numbers.) To consider an example of a slightly different sort, the communication system employed by honeybees also contains an infinite number of possible expressions. But the sense of the word "infinite" that applies to that system is not of the correct type for language, since it is continuous rather than composed of discrete elements. Thus Chomsky says (continuing a previously quoted passage from 1988, p.169):

> At this point one can only speculate, but it is possible that the number faculty developed as a by-product of the language faculty. The latter has features that are quite unusual, perhaps unique in the biological world. In technical

> terms it has the property of "discrete infinity." To put it simply, each sentence has a fixed number of words: one, two, three, forty-seven, ninety-three, etc. and there is no limit in principle to how many words the sentence may contain. Other systems known in the animal world are quite different. Thus the system of ape calls is finite; there are a fixed number, say, forty. The so-called bee language, on the other hand, is infinite, but it is not discrete. A bee signals the distance of a flower from the hive by some form of motion; the greater the distance, the more the motion. Between any two signals there is in principle another, signaling a distance in between the first two, and this continues down to the ability to discriminate. One might argue that this system is even "richer" than human language, because it contains "more signals" in a certain mathematically well-defined sense.[18] But this is meaningless. It is simply a different system, with an entirely different basis. To call it a "language" is simply to use a misleading metaphor.

If, analogous to the case of Helen Keller, we suppose that on a certain day tens of thousands of years ago, one or more sapiens who previously had no power to speak managed to learn just a single meaningful word, then that would have put those sapiens in a position, at least in principle, from which they could obtain (and could lead other people to obtain) a whole language as well, along with that one word. Of course, in some respects, the example of Keller is misleading for our purposes. For example, none of our pre-historic ancestors had a teacher who already possessed a language, and who therefore was able to guide his or her students towards rediscovering that same institution or phenomenon for themselves. To be still more specific, our early ancestors did not have any language-using chief, boss, seer, or priest who could manipulate, help, force or persuade other tribe members to acquire language for themselves. Furthermore, as mentioned earlier, even previous to the time when Helen Keller first met her teacher, she had invented for herself, or somehow had learned from her contact with the language-using people around her, quite a few tactile and bodily signs that were similar to language. For instance, Helen learned to affirm and accept some things proposed to her, and to deny and reject others, by placing people's hands on her head, and then nodding up and down for "Yes" and shaking her head from side to side for "No." Wild chimpanzees do not and presumably cannot employ signs of that sort for communicating with one another; and it is presumably true to say that our forebears who lived before

---

18 Let me (D.M.J.) propose a very simple example to make the meaning of non-discrete infinity still clearer. A bee's dance can indicate that the direction in which some newly found patch of clover lies, relative to the hive is NW, or NNW, or NNNW, or NNNNW, or NNNNNW, and so on, theoretically without limit.

the time of the UPR must have been in approximately the same position as chimpanzees.

Two sorts of considerations are relevant to what we are talking about here: (a) the background, preparation, and presuppositions for humans' acquisition of language; and (b) the particular event or series of events that distilled and somehow triggered explicit language out of that background. Keller's preparation for acquiring the meaningful word "water," included her previous, repetitive practice with hand-signs (directed by her teacher), including the sign for water. A second background factor was a vague memory of having spoken that same word, and having at least partially understood it, when she still was a sighted and hearing child. Still a third preparatory factor was the emotional incident of her breaking the unfamiliar doll by throwing it on the floor, as a protest against her teacher's attempt to get her to accept it as another, additional referent for the same one hand-sign, "d-o-l-l." Analogously, I claimed in the previous chapter that part of pre-linguistic humans' preparation for their acquisition of language might have been their domestication of animals like dogs, because experience with those creatures gave them practice in setting aside instinctual inclinations inherited from their mammalian and primate ancestors, and helped them develop new inclinations to take the place of the old ones. Nevertheless, the main thing I now want to discuss in what remains of this chapter is the second issue—i.e. the eliciting cause for humans' becoming able to learn at least one word—because it seems to me that enough has been said for the time being about preparations. The thesis I shall defend here is that the event that put an end to early sapiens' wolf- and raven-like existence may have been connected with what psychologists call "flashbulb memory."

Let me give a few examples of that to which I refer. The classic Arabian story, *Ali Baba and the Forty Thieves*, contains several insightful observations about how human memory works. This story focuses on two brothers, Cassim and Ali Baba, who inherited equal amounts of money and goods from their father, but who eventually had quite different fates. At first Cassim was the more successful of the two, because he acquired a fortune by marrying a wealthy widow, while Ali Baba married a woman as poor as himself, and finally was reduced to earning a meager living by gathering and selling firewood from a nearby forest. However, when working in the forest one day, Ali Baba had the good (albeit dangerous) fortune to overhear, from his hiding place in the branches of a nearby tree, the preparations a band of caravan robbers were making to enter the cave-like grotto where they and their predecessors had hidden stolen goods over the course of several generations. The most important thing Ali Baba heard during that time—all the while fearing for his life, lest he be discovered—was the voice of the captain of the thieves pronouncing the password that opened the grotto's door: "Open Sesame." After the thieves had finished their business and left, Ali Baba's knowledge of that password allowed him to enter the cave and take away some of the treasure. The following day, Cassim also learned this same password from Ali Baba, by threatening to denounce Ali Baba to legal authorities, unless he revealed it. Cassim then immediately set out to find the grotto himself, taking along 10 mules loaded with large empty chests, which he intended to use to carry away a much greater share of the thieves' booty than Ali Baba had taken. But when Cassim entered the thieves' cave, and looked at the many marvelous things it contained, he soon forgot the password. After that, he fell into a panic in which, the harder he tried to remember the words, the more impossible it became for him to do so. The next day the thieves returned to the grotto, found Cassim trapped inside, and killed him. By contrast, Ali Baba never had any trouble remembering the password, to the end of his long and (eventually) fortunate life.[19] What allowed him to remember the password, while Cassim could not do so? The answer, I suggest, is that the circumstances in which Ali Baba first heard those words were so unusual, dramatic, remarkable, and fearful, that they had the effect of creating an indelible memory.

Next, if someone does something I wanted to do myself but never succeeded in doing, or if he or she breaks a piece of news to other people that I was looking forward to telling them, before I have a chance to do that, I might say of this person that he has "stolen my thunder." What is the origin of this odd but very familiar English phrase? Using the internet, I found the following short account of it.[20] The saying, "steal my thunder" was made famous by the English critic and playwright, John Dennis, in the year 1709, because of the following situation. Dennis had invented a new way of simulating the sound of thunder on stage and had used that method in one of his plays, *Appius and Virginia*. Dennis "made" thunder by using troughs

[19] It never is explained in the story why neither Ali Baba nor Cassim wrote the password down, or (if they happened to be illiterate) why they did not devise some other visual reminder of it, like carrying a small bag full of sesame seeds on a string around the neck.

[20] http://www.trivia-library.com/b/origins-of sayings-steal-my-thunder.htm

of wood with stops in them instead of the large mustard bowls that had been employed before. Although the thunder was a great success, Dennis's play was a dismal failure. The manager at Drury Lane, where the play was performed, canceled its run after only a few performances. A short time later, Dennis returned to Drury Lane to see a performance of Shakespeare's *Macbeth*. As he sat in the pit, he was horrified to discover that his method of making thunder was being used. Jumping to his feet, Dennis screamed at the audience, "That's my thunder, by God! The villains will not play my play but they steal my thunder."

Since this incident was spontaneous and unplanned, no one had the slightest idea about the profound effect it eventually was going to have on English ways of speaking. Nevertheless, one main factor that helps to explain the fact that its consequences turned out to be surprisingly powerful was the unique and emotionally charged situation in which a large number of people first heard that phrase spoken, at the same time.

A possibly similar case is the insulting English expression, "old bag," that some people apply to any mature woman who happens to be romantically interested in a younger man. For many years, I carried around in my head the following couplet: "If my love hath a bag of gold, what care I if the bag be old?" I assumed during all that time that Shakespeare was the author of that couplet. (I certainly am not clever or poetic enough to have dreamed it up myself.) Furthermore, I assumed that this sentence had served English speakers as the flashbulb-memory-like source of the phrase just mentioned. But now I no longer am able to defend the idea that those two claims are true, since I could not confirm the truth of either of them in the course of the searching and inquiring I did to prepare this chapter. Nevertheless, I did not have the heart to leave this example out of the chapter entirely, for three reasons. The first (i) is that I love it so much; the second (ii) is that "old bag" sounds like the kind of thing that is likely to have started in the way we are discussing; and the third (iii) is that the book you now are reading is neither about Shakespeare, nor about the English language in particular; and therefore its author feels he ought not to be required to be rigorously accurate about exactly which poet or dramatist wrote or said what, where, when, and under what circumstances.

The last example I want to mention—which, happily, I do know something about—is the furor that surrounded the death of the then president of the United States, Franklin Delano Roosevelt, on April

12, 1945. At the time Roosevelt died, the enormous armed conflict of World War II, in which he had played a major role, was just coming to an end. Nevertheless, many people on the Allied side still felt a great deal of anxiety and fear about bringing the war to a quick, successful, and honorable conclusion, and felt that "this was no time for Roosevelt to die." As a result, all the people just mentioned were nervously unsettled and saddened, while some other individuals (prominently including enemy combatants) were elated, to receive the news that the leader of the Allies suddenly was gone. At that time, for instance, Adolph Hitler, the leader of the enemy state of Germany, was living from day to day in an underground bunker in Berlin, taking whatever desperate measures he could muster to avoid total defeat. When he received the news of Roosevelt's death, he experienced a brief period of renewed hope, in which he started to believe once more that (as in the career of his hero, Frederick the Great) events might be coming together to save him at the last minute. Similarly, Japanese-American soldiers fighting with the United States army in Germany did not choose to remember Roosevelt as the man who had signed the document that had put their families into internment camps, but instead proposed to honor him as the person who had given them a chance to prove their loyalty by enlisting in the country's armed services. When these soldiers were told about Roosevelt's death, they set out to pay him homage by spontaneously gathering up their weapons and going out to attack German positions, without having received any orders to do so, without having any coherent strategy or plan, and without having much regard for their own safety. Thus, quite a few people all around the world were able to say exactly where they were and what they were doing when they received the news of his death. And—immediately following the announcement—those same people found it difficult even to remember the name of Roosevelt's successor, Harry S. Truman.

I was too young to be a full participant in all this, since I was born in May of 1939, four months before the war began in September of that same year. But even in my case, there is a small story to tell. I was a five-year old child living with my father, mother, and baby sister in Ogden, Utah, when Roosevelt died. My mother, like many others, was very impressed and saddened when she heard about his death on the radio and read about it in the newspaper. (My father was temporarily absent because of his job.) I clearly remember that occasion, almost certainly because of the part my mother played in it. She looked into my eyes, took me by the shoulders and shook me hard, saying, "David, I want you to remember this day!" I recall being embarrassed by all of this, because it struck me as silly, undignified, and completely unnecessary. Therefore I repeatedly assured her that I would remember it if she wanted me to do so, and that there was no need for her to make this great fuss to help me remember. But of course, all of those assurances almost certainly were false, since it is now clear to me that I would not have had any memory of the day Roosevelt died, if she had not done what she did.

If Chomsky is allowed to make up fanciful and speculative but explanatory tales in the indirect service of science, I claim the right to do so as well. Here is my story. Suppose that 60,000 years ago, before any humans had learned to speak in a syntactical fashion, there was a tribe of sapiens living in a mountainous area, whose members were involved (like the forces of Hitler at the time mentioned before) in an increasingly desperate and dispiriting war against a strong, aggressive, and determined enemy—i.e. a neighboring tribe from a nearby valley. The people of the mountain tribe, like virtually all other humans then alive, were frightened and baffled by the phenomenon of lightning. Relatively speaking, however, they had become used to it, because they lived in a place where thunderstorms were common. On the other hand, despite the fact that all the warriors from the valley tribe who now surrounded their settlement were courageous and fearsome, they also were skittish about any lightning strikes that happened to land close to them, since they were not used to seeing and hearing lightning except from a safe distance. Then, at a culminating moment of the battle, when the valley people almost had given up hope that they could avoid being systematically slaughtered, their leader—a loved and respected, Roosevelt-like man of great experience, strength, and moral authority—spontaneously yelped out the sound "Leet!" at the moment when he was startled and killed by a shaft of lightning that struck his front-line position at a place close to the main thrust of the attackers. The loss of this leader gave the members of the mountain tribe still another reason to be discouraged. Nevertheless, the nearby members of the attacking tribe from the valley were far more terrified by the shock, power, and noise of the lethal lightning strike they had witnessed, than any of the defenders were. At that moment, in fact, a wave of panic began to spread through the ranks of the attackers, so that all of them finally were transformed into a frightened and dispirited mob who turned their backs and ran down from the mountain, never to return, leaving the defenders to shout out their relief and joy, and to weep with gratitude for their unexpected and dramatic salvation.

I hypothesize that, from that day forward, this "flashbulb" incident that had made such a strong and lasting impression on them, was sufficient to establish the sound "leet" among the mountain people as an objectively correct and universally accepted word, that did not just refer to the one stroke of lightning they witnessed that day, but referred instead to all lightning, wherever and whenever it might occur. For example, they might have thought of that spoken sound as a general "name" that was revealed to them by the spirit, demon, or god who sent the gift of lightning to rescue them at the moment of their greatest vulnerability. Furthermore, analogous to the experience of Helen Keller, I propose that almost immediately after the mountain people had obtained this one word by means of what they considered to be a divine revelation, they also then began to learn—in the sense of spontaneously inventing—many other words as well. In fact, it soon became clear that the number of words they were in the process of inventing was not just large, but was potentially inexhaustible. In this fashion, then, they might have become the first creatures on Earth to possess a syntactically organized language.

Some readers will complain that the preceding story is both unrealistic and unnecessary, because nearly all the young children we encounter today easily can acquire a language of this type, without their first having had to go through a complicated series of events like the ones just described. My answer to this objection is to remind critics that I am talking about members of our species who lived long ago, not people who are alive now. As noted in section 3.2, it does not seem to be the case that humans' evolutionary development slowed or stopped since the time of the UPR. Therefore we can plausibly assume that the genetic make-up that people have today is different from what it was 60,000 years ago. To mention a couple of simple examples, some scientists say that humans' biological nature almost certainly has changed along with the social conditions in which they live, so that people now are less physically suited to climb trees, and more physically suited to live together in large cities, than their ancestors were. Furthermore, since almost all the cultural groups into which humans are born today are literally saturated with language, it also should not be surprising to us that our children (gradually) have become genetically fitted so as to be able to learn language spontaneously, quickly, and without difficulty. Nevertheless, the hard question—to which theorists like Pinker and Chomsky do not pay much attention—is how this now familiar situation originated in the first place. My answer to that question is that individual humans did not begin, nor in fact, could they possibly have begun, to speak simply by willing to do so. Instead, this might have happened, in each of the several separate instances of the occurrence of the UPR, because of one or another emotionally charged, perhaps apparently divinely ordained connection of ideas that happened to be formed at the same time by some relatively large group of people, all of whom were members of the same cultural tradition.

Chapter 5

A Third, Even Earlier Invention that shaped Our Nature: Religious or Objective Consciousness

When I see something that makes absolutely no sense whatever, I figure there must be a damn good reason for it.

Peter De Vries

5.1 The reason religious thinking became universal for the members of our species was that it was a mode of thought that helped us see and understand things as they actually were

This is the everlasting and triumphant mystical tradition, hardly altered by differences of clime or creed. In Hinduism, in Neoplatonism, in Sufism, in Christian mysticism, in Whitmanism, we find the same recurring note, so that there is about mystical utterances an eternal unanimity which ought to make a critic stop and think, and which brings it about that the mystical classics have, as has been said, neither birthday nor native land. Perpetually telling of the unity of man with God, *their speech antedates languages,* and they do not grow old.

William James, *The Varieties of Religious Experience*
[italics added]

. . . the condition of a unity between thought and religion . . . overcomes their, so to speak, schizophrenic cleavages in personal and cultural life.

Paul Tillich

As far as observers (including a large number of well informed and scientifically trained investigators) have been able to tell, virtually all adult, undamaged, present-day humans, as well as many generations of their ancestors, do have at the present time, and did have in the past, a deep connection with the concept, practice, or outlook of religion. For example, Walter Burkert says on the first page of his (1996):

The observation that practically all tribes, states, and cities have some form of religion has been made repeatedly, ever since Herodotus. . . . [I]t is the universality of the consensus that has to be explained. . . . [Furthermore, t]he ubiquity of religion is matched by its persistence through the millennia.

> It evidently has survived most drastic social and economic changes: the Neolithic revolution, the urban revolution, and even the industrial revolution. If religion ever was invented, it has managed to infiltrate practically all varieties of human cultures; in the course of history, however, religion has never been demonstrably reinvented but has always been there, carried on from generation to generation since time immemorial.

What is the human universal that, in the opinion of Burkert, calls out for explanation in such a clear and obvious fashion? One would not be justified in simply calling it "religion," in the way Burkert himself does, if only because of the fact that there are a great many individuals who now describe themselves as agnostics or atheists, and who affirm, in an apparently sincere fashion, that they do not possess any such thing as religion. Nevertheless, it seems to me that there is something at least approximately correct about Burkert's way of talking, since we know from experience that all people who are and were more or less like us—including atheists and agnostics—habitually conceive of themselves and the world around them in quasi-religious terms. In this chapter, therefore, I shall content myself with talking, not about the supposed universality of religion, but merely about the universality among all of human beings like us, of a certain type of thinking that I shall call "religious consciousness."

Burkert makes the further claim that people have possessed religion "since time immemorial." But this does not seem to me to be a correct idea, because I do not believe it is appropriate to suppose that either religion or religious thinking is a biologically given feature of *Homo sapiens*, like the facts that we have two arms, that we grow flat fingernails rather than claws, and that the males of our species have larger penises than all other known primates. Instead, it is more plausible to suppose that a religious way of conceiving of things is an invention that some members of our species made in prehistoric times—probably more than once, at several different places and dates. Still more explicitly, the thesis I shall defend in the following pages is that religious thought was a once novel idea that one or several sapiens thought of approximately between the dates of 100 to 70 thousand years ago, and which, following that, gradually spread to virtually all the other members of our species, by means of learning, imitation, ritual, and other forms of cultural transmission.

Even if I am right to claim that religious consciousness is something that is basically cultural rather than biological in character, I believe it managed to persist and expand since the time period just mentioned, for an ordinary, Darwinian reason—namely, that it was advantageous for human survival. Furthermore, I shall argue that religious consciousness was useful to humans in a much more direct and basic sense of this word than most recent commentators concerned about such matters suppose.

Let me begin by talking about a case that sheds light on what I take religious consciousness to be. This case is a series of things that happened in the life of the seventeenth-century Dutch-Jewish philosopher, Benedict Spinoza. This man's reflections on topics in philosophy, ethics, history, and biblical interpretation eventually led him to conclude that "God" and "the world" were two names for (i.e. two different ways of looking at or thinking about) the very same thing. In particular, Spinoza was impressed by several passages he found in the Old Testament where (i) various events seemed to have happened completely as a result of ordinary, accidental, and physical causes, but (ii) God later revealed, through His prophets, that He himself had brought about those events, in the sense that He had foreseen them, and had made them to happen, as a result of His will, intentions, and planning.

However, as soon as Spinoza let it be known that he believed God was the same thing as the world, some members of his religious community gave him the label of a subversive and dangerous atheist. Following that, those same people began to shun him, then actively hounded him, and finally managed to have him excommunicated from his synagogue. However, the actions of those self-righteous individuals not only strike me as having been cruel and unjust, but also were positively misguided in my opinion, because of the following point. All the people who knew Spinoza best—i.e. the ones who were most closely associated with him, and who therefore were most familiar with the contents of his mind—reported that he was not at all opposed to ideas and ideals like faith in and reverence towards God. In fact, those people said, Spinoza accepted a fairly standard distinction between human virtue on one side and sin on the other, and frequently expressed approval of the first and disapproval of the second. Again, they said he consistently believed in and supported such things as the sanctity of prayer, worship, and piety. In fact, instead of hating and rejecting everything of the kind just mentioned, his friends described him as a humble and worshipful man who was "intoxicated with religion."

Thus, as opposed to the opinions of people like Spinoza's detractors, I claim it is more illuminating and accurate to identify religious belief with a certain mental attitude or habit of thinking, rather than with an affirmation of some vague, obscure, and problematic proposition about a being or group of beings that exist in some unknown parts of our universe. For example, it does not strike me as something that is either true or meaningful to claim that the one essential thing presupposed by religious piety is that a person should posit the existence of a personal, all-powerful, all-knowing, "supreme being" who, in one or another mysterious fashion, was directly responsible for the creation of everything (else).

To be still more explicit, I conceive of religious consciousness as the power and practice of considering and contemplating oneself and all other things, in a way that is detached, serious, full of awe, (usually but not always) optimistic, and relatively selfless.[1] For example, it seems to me that the following quotation from Thomas Merton—in spite of the fact that it includes words like "God," "sin," and "truth"—expresses roughly the same idea about the nature of religion and a religious mentality as the one that I also accept.

> At the center of our being is a point of nothingness which is untouched by sin and by illusion, a point of pure truth, a point or spark which belongs entirely to God, which is never at our disposal, from which God disposes of our lives, which is inaccessible to the fantasies of our mind or the brutalities of our will. (Merton, 1966b, p.158.)

Let me now give a quick summary of the general project in which I am engaged. Much of the first part of the book I published in 2003 was devoted to a discussion of the pre-historical occurrence that many specialists call "the Upper Paleolithic Revolution." My expanded aim in this second authored book (as indicated by its title) is to talk not just about one, but about three pre-historical inventions that were important cultural sources of present-day human nature. The Upper Paleolithic Revolution (along with the acquisition of complex language that presumably was the cause or trigger for it) is the third and latest in time of the three of the inventions discussed in this book. The two earlier ones were, the invention of religious consciousness, and following that, the first taming and domestication

---

[1] In this respect, I take religion to be somewhat analogous to my own professional area of philosophy. The way I often introduce beginning students to this field is by telling them that philosophy is not a traditional intellectual discipline like chemistry, zoology, or economics whose job it is to pick out some specific set or kind of things that people can think about. On the contrary, it is a method of thinking that each person is free to apply (sometimes usefully, sometimes not) to nearly any object, topic, concern, or problem that he or she might choose.

of animals. The two events just mentioned were, in my opinion, changes that helped to prepare the members of our species for their eventual acquisition of full language and the UPR. Thus, I claim that the beginning of the long and laborious historical process by which our ancestors succeeded in separating themselves from other animals, including all the other hominids, began at the time when (probably about 100 to 70 thousand years ago) some of them became explicitly aware of a previously unexploited power to engage in religious thinking. This power presumably was something that had lain dormant, ignored, and inaccessible since the very beginning of our species.

Accordingly then, my two authored books, considered together, tell a story about four cultural steps our ancestors took on the way to developing the human nature we now possess. Those steps were—in order of occurrence—(i) the invention of religious thinking, followed by (ii) the first domestication of animals, followed again by (iii) the invention and exploitation of sophisticated language, then followed finally by (iv) the pre-Classical Greeks' invention of reason around the year of 1500 B.C., about which I spoke at length in my previous book.

Returning now to previous concerns, we can say that although religious thinking is something that belongs to all undamaged present-day humans, it does not apply either to creatures of other species, or even to those members of our own species who lived before the date of approximately 100 thousand years ago. This is true because other species of animals do not participate in, and *a fortiori* do not count as parts of, human culture. Similarly, it also does not apply to our ancestors of more than 100 thousand years ago, because those ancestors lived either before that invention had been made, or before it had enough time to spread to virtually all humans.

If all this is approximately on the right track, then it must be a mistake for atheistic thinkers like Richard Dawkins and Daniel Dennett (or "brights," as they like to call themselves) to say that religion always undermines and spoils a person's ability to think in clear and dispassionate terms. On the contrary, far from being something that hindered, deflected, and destroyed such an objective conception of things, it seems to me that the invention of consciousness of that type was the first and most significant step humans took on the way towards thinking of that type. I suggest it might be a better idea to interpret the critical remarks Dawkins and Dennett make about "religion," as being implicitly aimed instead at the arrogance, tyranny, fanaticism, and other abuses into which some members of organized religious sects occasionally have fallen at certain times in the past. The reason this is true is that fanaticism of that sort undoubtedly does undermine people's ability to think in a clear, reasoned, and objective manner. However, the authors just mentioned fail to take account of the fact that many intelligent and sensitive people have dedicated their lives to trying to expose and reform abuses of that sort, without thereby losing their own religious convictions as a result of doing that. In fact, I believe that fanaticism and the other forms of arrogance just mentioned deserve to be considered the antithesis of civilized, thoughtful, and sophisticated religion, rather than its center and essence, since fanatics are people who make ridiculous and ultimately fruitless attempts to "play God," rather than following the practice more typical of religious thinking, of emptying themselves in such a way as to accept a religious interpretation of the world and humans' place in it.

Since religious consciousness, as I interpret it, is a conception of oneself as being separated both from God on one side, and from the natural world on the other, thinking of that sort is general rather than specific, and is most correctly conceived in negative rather than positive terms. To say the same thing another way, the crucial thing a religious person believes is that he or she is *not* God, and furthermore that he is neither a part of, nor the creator of, nor a dictator to, the things, facts, and properties he observes in the world around him. Why do I believe that conceiving of matters this way is that which enabled humans to take their first steps into the realm of objective and rational thinking? The answer is that the most fundamental thing required by mental activity of that sort is that a person should draw a fairly sharp distinction between himself or herself and all the objects of his attention.[2]

---

[2] John Buchan, a novelist and former Governor-General of Canada was quoted in my newspaper as follows:

> Religion is born when we accept the ultimate frustration of mere human effort, and at the same time realize the strength which comes from union with superhuman reality.

I agree with the general thrust of this statement. But I also think a couple of changes need to be made in it. First, instead of talking about the "frustration of mere human effort," it is more accurate to speak of the weakness and inefficiency of trying to understand matters by means of thought that is engaged, prejudiced, self-interested, and (in this respect) animal-like. Second, instead of talking about the strength that

Let us now turn to the question of how religion managed to survive and grow, until at last it became nearly universal among humans. I think this happened for roughly the same reason that applies to other, more familiar inventions that also have become widespread. For example, why does nearly every person in today's world aspire to have a telephone, a computer connected to the internet, modern medical care, and an effective sewage system? The answer is that all of those items are widely—and correctly—considered to be useful tools for living a healthy, happy, sane, and productive life. So in similar fashion, I believe the reason it no longer is possible to find any tribe or group of human beings on Earth that does not possess some kind of religion[3] is that the sort of thinking in which religion allows people to engage, has proved to be so helpful that no sensible and rational person who is not affected by various forms of prejudice would want to be without it.

What is the main adaptive advantage religious thinking has for humans like us? As a starting point for answering that question, consider some alternatives to the view I have been proposing, in the form of a short but typical list of suggested Darwinian advantages of religion put forward by other investigators.

(a) Religion contributes to a person's health, happiness, and well-being. (Plato) It is a tonic. (William James)

(b) Religion is that which holds together certain kinds of groups of people, and provides members of those assemblages with a quick and easy way of recognizing, isolating, and thereby punishing any individuals who are not genuinely interested in the welfare of those groups.

(c) Religion tends to give the people who have it a good reputation; and it is an effective means by which they are able to gain the trust of others.

---

comes from a "union with superhuman reality," it strikes me as more realistic to speak of the strength that is associated with thinking that is humble, respectful, and disengaged from all selfish interests and desires.

[3] I have learned the hard way to be suspicious of "all" claims like this one. Will there come a time when I will have to take back this claim, by writing yet another book partly dedicated to that purpose? In one obvious sense, I hope the answer is no. Nevertheless, I am more concerned with trying to discover true things to say about present-day humans and their nature, than I am with avoiding embarrassment and preserving my own comfort.

(d) Religion makes friendship—and thus also marriage, breeding, and reproduction—easier, more robust, and more assured.[4]

I do not deny that there is a great deal of commonsense truth contained in the points in the preceding list, as well as in similar ones that might have been added to them. Nevertheless, a problem is associated with all proposals of this type—the problem, namely, that there is at least a verbal conflict, and perhaps also a genuine contradiction, between each of the items on the list, and the observed fact that religion is universal among modern humans. To express the same point another way, if it really were the case that people came to accept religion for reasons like the ones just mentioned, then religion would not have been sufficiently fundamental to have become incorporated into human nature itself. For example, if religion only were useful for specific, detailed, and practical reasons like those included in the previous list, then—contrary to what we observe—anthropologists occasionally would have found isolated groups of humans to whom religion was unknown (again, see footnote 3).

Thus, a kind of relativity is present here. It is possible to distinguish (as Dawkins and Dennett do) between those humans who are comparatively more friendly to, and less embarrassed by, religious ideas like God, sin, repentance, resurrection, and life after death, and who therefore are inclined to talk and think explicitly about such matters; and other individuals who are not like this, because they have different personalities, preferences, and priorities. But although distinguishing between people of those two types might be useful for various purposes of ordinary life, I do not believe that same distinction can help us understand the nature of religion, considered in general. One reason this is true is that the above distinction does not throw light on the question of what all the undamaged members of our species have in common.

Thus—if readers will forgive me for repeating something said before—I claim that some of our ancestors invented the practice or institution of religious-like thinking, when they started to conceive of themselves as different from God and from the rest of nature. After that, however, it gradually became clear that this way of thinking did not have very much to do with obscure, complicated, and controversial theological questions like whether there exists a God who is sufficiently powerful to grant us life after death. Instead, it was more

---

[4] Points (b), (c), and (d) are mentioned in the article in *The Economist* listed in the bibliography under the heading: "Anonymous. 2008."

usefully applied to matters that were more limited, more concrete, and more practical. To be precise, the principal usefulness of religious-like thinking was that it gave people a power to think more clearly about any of the things they met in their experience, because it enabled them to focus attention on the objects in which they happened to be interested, without confusing and mixing those objects together with personal and emotional considerations like whether they approved of the items in question, or desired them, or whether those items made them feel good, and so on. Further, thinking in this non-involved, "hands-off" manner also helped people to obtain a greater amount of control over the world (compared to their primate relations, and the earlier members of their own species), because it allowed them to understand what the world itself was like in a sharper and more accurate fashion. Stated still more simply, the main adaptive value of religious thinking was that it gave people the power to contemplate all the various things around them, in ways that were comparatively free from self-interest.

5.2 The extinction of the Neanderthals, and other trophy wars

> It is hard to tell, but *something like* religion may well have existed from the early days of language . . . or even before that. What were our ancestors like before there was anything like religion? Were they like chimpanzees? What, if anything, did they talk about . . .?
>
> Daniel Dennett, 2006, p.102

A sure sign that humans had developed the power to domesticate animals, which I discussed in Chapter 3, is that remains of those animals regularly started to appear in graves and other deposits along with the bones of humans. Furthermore, the skeletons of the first creatures humans domesticated—presumably wolves/dogs—began to change in various respects (e.g. in their overall size, in the shape of some of the bones, in the internal composition of almost all the bones, etc.), compared with the bodies of their wild ancestors. Similarly, one sign that humans had acquired sophisticated language—the topic of Chapter 4—was the extensive suite of changes that were associated with the Upper Paleolithic Revolution, in all of which language apparently played a pivotal role. As mentioned before, for instance, humans' newly developed facility with words suddenly made them able to produce carefully distinguished tools of many more kinds than once had been available to them, including multi-piece tools like spears with stone heads affixed to them, and spear-throwers that enabled spears to travel farther, as well as tools constructed out of materials other than stone, like bone needles and awls. This also allowed them to create the first artistic objects like pictures, musical instruments, and items of personal adornment like beads. Furthermore, at more or less the same time, people invented ocean-going boats, began to engage in long-distance trade, and then started to colonize previously uninhabited islands as well as the continent of Australia-New Guinea.

Analogously, if what I am talking about in the present chapter is not just a figment of my imagination, it seems reasonable to believe that there also should have been one or more observable signs that were associated with the slightly earlier hypothetical event I propose to call the beginning of religious-like thinking. My project in this second section of the chapter will be to argue that there was at least one such sign—a sign that was closely connected with what Ian Tattersall refers to as the mysterious extinction of the Neanderthals.

It is a basic principle of evolutionary biology that, sooner or later, all species are bound to become extinct. Thus, what leads Tattersall to consider the extinction of this particular species mysterious?[5] Part of the answer is that although the vast majority of extinctions before that time happened gradually and over a comparatively long period, this one was remarkably rapid. Furthermore, as far as scientists can tell, it did not happen in conjunction with any destructive natural event like a flood, a fast spreading disease, a volcanic eruption, or an asteroid strike. Instead, it seemed to have been a byproduct of certain interchanges between Neanderthals on one side and members of our own species on the other.

Let me try to put this last point into historical perspective. At the time our species came into existence (evidently in Africa), and then later started to move out of this continent of origin, there were at least two other widespread species of hominids that already were present in the rest of the Old World. One was a late form of our "grandfather" species, *Homo erectus*, which inhabited East Asia. (Two names earlier archeologists once gave to that eastern species were "Java Man" and "Peking Man.") The other was a "cousin" species that was more

[5] The full title of the book I am talking about here, whose revised edition Tattersall published in 1999, is *The Last Neanderthal: The Rise, Success, and Mysterious Extinction of Our Closest Human Relatives.*

closely related to us, to which we now refer as *Homo neanderthalensis*, which occurred in the Middle East and Europe, where it also may have originated. Although our sapiens ancestors were not in close physical contact with members of *Homo erectus* in East Asia (at least in comparatively early times), they did live for a long while in intimate proximity with the Neanderthals who, as just was said, were much more similar to us in various respects than *Homo erectus* was. For example, we already noted that, during roughly half the time our species has existed, our ancestors lived in ways that were virtually indistinguishable from the life style of their Neanderthal neighbors. (In addition to the passage from Thorne and Wolpoff, 1992, summarized before in section 2.2, further information about this point can be found in Ian Tattersall 1998, pp.150-80, and 1999, Chapter 8.)

However, this picture changed some time between the dates of 100 thousand to 70 thousand years ago. The reason things changed during this period was that this was when the life-styles of *Homo sapiens* and *Homo neanderthalensis* began to diverge. I believe the root cause of this divergence lay in the fact that some members of *Homo sapiens* had began to think about things in a new way, but the Neanderthals were not also able to adopt this same, special way of thinking. Thus, I now propose that the observable, historical mark of the beginning of religious consciousness is the beginning of the mental divergence between our species and the species of the Neanderthals—a divergence that eventually would lead to a systematic and consciously directed program whereby one of these species was completely extinguished by some members of the other species.

I am inclined to accept the claims of paleoanthropologists like Tattersall who argue that, in spite of the fact that the Neanderthals were more closely related to us than any other creatures, only humans of our species were able to engage in religious thinking, and not the members of any of species of hominids, including the Neanderthals. Tattersall says, for example, that Neanderthals probably were motivated to wear clothes, to bury their dead, and to help their aged and sick relations, not because of some religious conception they had of themselves and the world in which they lived (e.g. the idea of their being special and separate, or of life after death, or of a powerful God who rewards virtue and punishes sin). Rather, they acted in those ways from immediately practical considerations like protecting themselves against cold; making sure that corpses of dead species members were not left lying around to cause disease or attract lions and hyenas into their living places; and enabling ill, injured, or dying tribe members to continue contributing to the welfare of their group for as long as they could do so.

Tattersall claims in one of his books (1998, pp.164-5) that, in the thirteen thousand years following the date of about 40 thousand years ago, the Neanderthals were edged out of their last (Spanish) refuge in Western Europe, and into extinction, by early modern humans who by that time had become equipped with the full cultural and technological panoply of the Upper Paleolithic, including sophisticated language. Even though this particular extinction was comparatively rapid, it was not an overnight event, even in specific places. Furthermore, towards the end of their existence, the Neanderthals took a few faltering steps in the direction of responding to some examples that had been provided by their post-UPR sapiens neighbors, and thus also in the direction of competing with those neighbors. At a few sites in western France and northern Spain, for example, there is evidence of a material culture known as the Châtelperronian, in the period between about 36 and 32 thousand years ago, which incorporated certain aspects of Upper Paleolithic technology (in particular, the production of multiple long, thin "blade" tools, notably burins, and a certain amount of work on bone). Experts now generally agree that this culture was associated with Neanderthals. For example, at Roc de Combe and La Piage in western France, the Châtelperronian alternates in the archaeological layers with the Aurignacian culture of the earliest moderns; and this provides a good reason for supposing that the replacement of the Neanderthals by moderns was not just a one-step process. For instance, in one Châtelperronian layer at Arcy-sur-Cure there was found (along with a Neanderthal fossil) a carved bone pendant, complex in shape and quite finely made and polished. Tattersall believes that this most likely was an item of personal adornment worn by some Neanderthal.

Nevertheless, the relatively unique object just mentioned was made at a time when modern human incursion into the area had already been achieved; and increasing numbers of archaeologists now believe that various aspects of the Châtelperronian were a result of the Neanderthals' copying, either directly or indirectly, the Upper Paleolithic technology they saw being employed by sapiens. A second, alternative suggestion is that objects of this sort might have been acquired by Neanderthals, by trading and bartering goods with modern humans. Either way, it is not plausible to suppose that the things in question were a Neanderthal invention. The Châtelperronian was a very late and terminal development out of the Mousterian.

Furthermore, as mentioned before, Tattersall and most other experts do not believe the Middle Paleolithic transformed itself into the Upper Paleolithic via the Châtelperronian, or that the physically different Neanderthals evolved into modern homo sapiens.

In Tattersall's opinion, then, it is not plausible to believe that the Châtelperronian blurred the distinction between Neanderthals on one side and modern humans on the other, either in the realm of technology or in the realm of symbolism. He says that everything we know about the Neanderthals points to the conclusion that, even though their brains were slightly larger than ours, they were cognitively different from us. Furthermore, according to him, that particular cognitive difference almost certainly played a role in their disappearance.

Tattersall remarks (ibid. p.166) that we should not think of the Neanderthals as simply an inferior version of *Homo sapiens*, because of the fact that there are a great many ways of playing the evolutionary game, and the Neanderthals played to a different set of rules than our ancestors did. With changing circumstances, he says, their rules eventually let them down—as ultimately our rules might do as well. But although that is true, comparing ourselves with the Neanderthals provides the best means we have of discovering and assessing the special sort of uniqueness our form of human nature has made available to us.

What ended the situation where the Neanderthals, who once preyed on other animals, and who even may have exterminated some species of those creatures in the process of doing this, were themselves exterminated by another, similar species of hominids that for a long time in the recent past had lived beside them peacefully? That extinction appears to be an incontestable fact of prehistory. The only areas of uncertainty associated with it are questions about how it is appropriate to describe it, account for it, and evaluate its significance in a wider context. As pointed out earlier, my own view is that the crucial divergence between the two species took the form of a comparatively small refinement in the kinds of thinking that *Homo sapiens* did, which eventually had a large effect on other matters—including the evolutionary fate of *Homo neanderthalis*. Furthermore, this refinement was something that only became available to the first of the two species just mentioned rather than the second, because of a certain biological potentiality that already existed in the brains of sapiens, but not in the brains of the Neanderthals.

I suggest that the thinking done by both Neanderthals and pre-divergence sapiens was superstitious in character, as indicated by the fact (to be explained in more detail in the following section) that it was based on gradual transformations of previously acquired habits. However, one main cause of the divergence that began to separate these two species was that some sapiens had gained access to an additional type of thinking that allowed them to make rapid, "chosen" transitions into new mental perspectives. To be still more explicit, the new form of thinking about which I now am speaking, which I propose to call religious (as opposed to superstitious) consciousness, was non-habitual in character. One of the effects this sort of thinking had was to give those members of our species a power to "target" their Neanderthal neighbors, in a way that finally would destroy those neighbors, along with any competition they might have posed. It is not plausible to suppose that thinking of this new type was just a matter of sapiens' gradually extending their already familiar tradition of hunting prey animals for food, to killing Neanderthals for that same purpose as well. I say this because eating hominids of any other species, would not have provided sapiens with a means of satisfying their hunger that was nearly as effective, natural, and safe, as the hunting of (say) horses and deer.[6]

I think the basic factor on which sapiens' newly acquired power over the Neanderthals depended was a simple principle of military tactics about which I learned many years ago, in a class on the subject of fencing with foils. The principle is this: If you can impose a rhythm of expectation on your opponent, then you already have gained the power to defeat him, because you then are in a position to break that

---

[6] Nevertheless, some evidence points to the idea that sapiens' eating of Neanderthals occasionally did take place. For example, a short note entitled "Eating the Competition?" recently appeared in my newspaper in which Robin McKie is quoted in London's *The Observer* as saying that a Neanderthal jawbone recently was discovered that apparently had been butchered by modern humans. Furthermore, the head of the research team working on this bone now believes that the flesh of the skull had been eaten by humans, and that its teeth may have been used to make a necklace.

Another short note ("Who's for Lunch?" *The Globe and Mail*, Monday, August 30, 2010, p.L6) reports that Spanish archeologists working on remains of the species, *Homo antecessor*, who inhabited Europe about one million years ago believe those hominids evidently did engage in the practice of cannibalism in order to satisfy their nutritional needs. (Those same archeologists make still another interesting claim—namely that *Homo antecessor* was the last known common ancestor between the African lineage that gave rise to our species, *Homo sapiens*, and the lineage that led to the European Neanderthals.)

rhythm unexpectedly. One can formulate that same point in a way that makes it fit in even better with the case at hand, as follows. If your opponent only can respond to challenges, in habitual as opposed to reasoned and chosen ways, while you have an additional ability to consider matters from a viewpoint based on direct examination of what you see, hear, and feel at the present moment, as opposed to the types of things you repeatedly experienced in the past, then you can exploit that difference for your own advantage.[7]

To summarize, then, on the assumption that Tattersall and other experts are right to say what they do about the points mentioned in the last few pages, there still is one more comment to make about this subject. The additional comment is this: The beginning of the eventually fatal cognitive split between the two species of hominids who lived together in Europe and the Middle East took place at a much earlier time than the actual act of extermination itself. Furthermore, we have some reason to speculate that this beginning point was when members of one of the species first started to think about themselves in religious terms, but the members of the other species were not able to respond to that challenge by doing the same thing.

Let me now try to throw light on what I take to be the physiological basis for the post-divergence gulf that developed between sapiens and Neanderthals (a difference that proved fatal for the second of those species), by paying attention to some empirical facts about how human brains are organized. As noted before, sapiens' and Neanderthals' brains were not very different in size. However, it presumably is true that—eventually—certain parts of the brains of those two species began to work in different ways. In particular, experts say that the prefrontal cortex of sapiens' brains is and was larger, and apparently also had a larger role to play, than the prefrontal cortex of the brains of any of the other primates, including the Neanderthals.[8]

---

[7] There is a story in Plutarch's "Life of Alexander" about Alexander the Great using much the same trick to overcome one of the most capable and stubborn opposing generals he ever had met. He commanded his besieging army to march up and down in front of his opponent's defensive position every morning, with blaring trumpets and the sound of swords being struck against shields. After that commotion went on regularly every day for more than a week, the enemy commander got used to it, concluded that it was nothing more than bluff, and no longer bothered to prepare his troops for an attack. But that was precise moment at which Alexander chose to launch his attack, which ultimately proved successful.

[8] Terrance Deacon says our prefrontal cortex is roughly twice as big as it would be, if it simply were the case that we had a relatively large ape brain. See the references to this point in Lieberman, 1991, p.101.

What is the characteristic function performed by the prefrontal cortex of the brains that belong to our species? (Or, to speak more correctly, what did this part of the brains of our ancestors finally become able to do, after the time of the divergence?)[9] The general answer to this question is that the prefrontal cortex is the part of the brain that allows us to learn new things, to look ahead, and to plan. For instance, Donald Stuss and Frank Benson (as quoted in Lieberman, 1991, pp.99-100) say the prefrontal cortex is the anatomical basis for the function of control. In other words, this part of our brains is crucially important at the time any new activity is being learned, so that conscious control of that activity is required. But once the activity in question has become routine, other brain areas are sufficient to handle it, and participation by the frontal cortex no longer is needed.

A few years ago, several students who took one of my philosophy classes, who happened to be majors in psychology, informed me of the following point. They said that "medical" movies like "One Flew Over the Cuckoo's Nest" inaccurately depicted what happened to a person after he or she had the operation commonly known as a frontal lobotomy (a surgical procedure doctors no longer can perform legally at the present time). The reason for this inaccuracy is that it is not true that this operation "turns the patient into a vegetable." According to these students, if you saw several people sitting around a restaurant table chatting informally, and one of them had received the operation called a frontal lobotomy, you would not be able to tell at first glance which one it was, because the behavioral effects of the operation are fairly subtle. The "tell-tale" signs are abnormal changes in someone's power to initiate, direct, and control things. Thus, suppose one of the individuals at the table says, "I have an idea! Let's go over to George's and see if he has any left-over pizza in his refrigerator." In that case, observers would be justified in concluding that the person who just spoke was not the one who had the lobotomy. But if someone else then says, "That's a great idea; I'm all for it!" informed observers would be in a position to infer that this second

[9] Again according to Philip Lieberman (see 1991, p.23), the claims made by nineteenth-century phrenologists were not completely mistaken, since observations and experiments of twentieth-century physiologists have shown the truth of the point that each of many areas of the human (e.g. *Homo sapiens*) cortex actually do perform various specialized and modality-specific jobs.

person might be (but of course, need not be) the one with the lobotomy.[10]

In effect, what a lobotomy does is to isolate the prefrontal cortex from the rest of the brain, so the remainder of the brain no longer can be influenced by that part. In a sense, then, people who have this operation lose their ability to make use of the most distinctive mental characteristic that belongs to undamaged members of our species, because in their case, that feature of their sapiens brains no longer works. Thus, rather than saying that Neanderthals were mindless brutes, I accept historical researchers' judgment that they were comparatively intelligent humans, who showed themselves to be successful, and in some respects even innovative. In spite of that, however, the mental means the Neanderthals employed to achieve their successes might have been something like the resources that remain available to any member of our species who has had a lobotomy. To make the same point in a different fashion, Neanderthals might have been able to obtain their new ideas by gradually transforming certain older, habitual ways of thinking, rather than (as normal homo sapiens often do today) by suddenly setting out in a new direction of thought.

Let us now consider a less speculative matter—namely, the historical observation that the extinction of the Neanderthals was only the first in a very long series of more or less similar historical events, in which our sapiens ancestors were involved. In particular, it is true to say that virtually every one of the places to which members of *Homo sapiens* traveled, after they left the Old World of Africa and Eurasia, also suffered an extinction of what biologists call its megafauna—i.e. its largest and most impressive species of animals—soon after their arrival. For example, Jared Diamond claims in his book, *Guns, Germs, and Steel* (1999, pp.42-3), that the settlement of Australia/New Guinea

---

[10] Lobotomy-like changes in the brain sometimes come about by "natural means," without the help of an operation. For instance, I once saw a television program about a law student who was sitting on a couch in his apartment calmly talking with this wife about everyday affairs, when he suddenly got a severe pain in his forehead. In the days and weeks following that incident, it gradually became clear both to him and his wife that he no longer could continue to be a student. Instead, he was forced to find a new kind of job, as a truck driver delivering packages to various addresses in the city where he lived. His employers in that job became aware of the fact that although he was able to carry out all tasks they assigned him, as long as they gave him clear directions and orders, he did not have the independence of mind, or a mental perspective of the right sort, to give orders to other people.

might have been associated with still another innovation, besides humans' first use of watercraft and their first major range extension since reaching Eurasia—namely, the first mass extermination of large animal species by humans. He says we think of Africa today as the continent of big mammals; and modern Eurasia also has many species of big mammals (though not in the same abundance as in Africa's Serengeti Plains), such as Asia's rhinos and elephants and tigers, and Europe's moose and bears and (until classical times) lions. By contrast, Australia/New Guinea has no equally large mammals at the present time. In fact, it has no mammal larger than 100-pound kangaroos. In the past, however, Australia/New Guinea had its own diverse group of big mammals, including giant kangaroos, cow-sized rhinollike marsupials called diprotodonts, and a marsupial "leopard." It also formerly had a 400-pound ostrichlike flightless bird, plus some impressively big reptiles, including a one-ton lizard, a giant python, and land-dwelling crocodiles.

Diamond tells us that all of those Australian/New Guinean giants (the so-called megafauna) disappeared after the arrival of humans. Scientists have disputed about the exact timing of their demise. But it is a revealing fact that several Australian archaeological sites, with dates extending over tens of thousands of years, and with abundant deposits of animal bones, have been carefully excavated and found to contain not a trace of the now extinct giants over the last 35,000 years. Thus it follows that the megafauna probably became extinct soon after humans reached Australia.

Diamond says the near-simultaneous disappearance of so many large species leads us to ask what caused this change. An obvious possible answer is that they were killed off or else eliminated in some indirect way by the first arriving humans. According to him, it is important to remember that Australian/New Guinean animals had evolved for millions of years in the absence of human hunters. Galápagos and Antarctic birds and mammals, which similarly evolved in the absence of humans, and did not see humans until modern times, still are incurably tame today. These last mentioned animals would have been exterminated, if conservationist authorities had not imposed protective measures quickly. On various other recently discovered islands where protective measures did not go into effect quickly, exterminations in fact did take place. For example, one such victim, the dodo of Mauritius, has become virtually a symbol for extinction. We also are aware of the fact that, on every one of the well-studied oceanic islands colonized in the prehistoric era, human colonization led to an extinction spasm whose victims included the moas of New Zealand, the giant lemurs of Madagascar, and the big flightless geese of Hawaii. Just as modern humans walked up to unafraid dodos and island seals and killed them, so presumably, Diamond says, prehistoric humans walked up to unafraid moas and giant lemurs and killed them too.

Hence a plausible explanation of the demise of Australia's and New Guinea's giants is to say that they met that same fate around 40,000 years ago. In contrast, most big mammals of Africa and Eurasia survived into modern times, because they had coevolved with proto-humans for hundreds of thousands or millions of years, and therefore had enough time to evolve a fear of humans, as our ancestors' initially poor hunting skills slowly improved. The dodo, moas, and perhaps the giants of Australia/New Guinea had the misfortune of suddenly being confronted, without any evolutionary preparation, by invading modern humans who possessed fully developed hunting skills. (On this same subject, also see E.O. Wilson, 1992, pp.249-53.)

What have these matters got to do with the thesis of this book? I submit that the historical events just mentioned were an expression of a certain kind of causal factor that previously had been unknown on Earth—an expression, namely, of what we might call the "monster" or "trophy" syndrome. This factor took the form of a conscious and systematic targeting of one general type of animals after another, based largely on something like a warped sort of pride. It was something like a way of saying "Here I am! Look at what I can do!"[11] (Prior to the time when sapiens began engaging in practices of this sort, even the Neanderthals may have had thoughts of a roughly similar sort, as shown for example, by quotations from an old book by George Constable *et al.*, 1973, pp.108-9, about a cave in Austria that was discovered to have a great many skulls of giant cave bears,

---

[11] A short, anonymous review article in my newspaper ("Homo unpleasantus?," *The Globe and Mail*, Thursday, February 15, 2007) quotes the primatologist Adrian Barnett writing in *The New Scientist* about his new book *The Last Human* as saying that, although most mammals have at least one or two close living relatives, we humans are unusual mammals since, along with our bipedalism, our naked skin, and our unusually large size, there is also the uncomfortable fact that we are the only living member of our genus. Furthermore, we modern humans progressively have become ever more isolated as we push, not only our closest relatives, the great apes, to the edge of extinction, but many other species as well. This action, he believes, is the latest manifestation of a violent intolerance of competition that has characterized the genus Homo, and probably also at least some of the other primate species that preceded our genus, for a long time.

arranged in such a way as to make all of them face towards the entrance, and dating from the time when Neanderthals were the only hominids in Europe.)

If what I have been saying so far in this chapter is not misleading, it might be correct to think of the events just described as early, paradoxical expressions of religious consciousness, or what I also have called the power to think in terms that are (relatively) objective. The paradox of which I now am speaking consists in this: Although many present-day people believe religion always and necessarily is associated with generosity, mercy, and deep unselfish love, it appears to be the case that the first and most primitive expressions of that sort of consciousness took the form of murderous hate. Still another way of talking about this paradox is to say that our early sapiens ancestors who first began to conceive of things in a religious manner had not yet learned the lesson of "letting God be God." Instead, like so many others through our species' long and troubled history, they fell into the error of trying to usurp God's place for themselves.

5.3 Superstitions are unconsciously formed reactions to patterns of experience that are based on unexamined wishes and fears; but religious consciousness is thinking of a more dispassionate sort, which can provide a rational basis for hope

I suggest that only relatively late, modern members of our species (as opposed to creatures of all other kinds) succeeded in acquiring religious thought in addition to superstition. What do I take to be the difference between the two things just mentioned? The answer—summarized in the title of this section—is that a superstition is an automatic, unconscious, and irrational "reaction of accommodation" that either a human or some other relatively complex creature makes to a pattern present in his or its experience. Furthermore, the basic cause of that reaction is something the creature in question wants to be true, or something it fears might be true. On the other hand, religious consciousness is a type of thought that is more considered and deliberate than superstition. In particular, one main reason religious thinking has turned out to be practically useful to us, is that it allows us to make a certain amount of progress towards understanding things objectively or, to say the same thing another way, towards a relatively unemotional and selfless comprehension of the world and our place in it. Even more particularly, religion differs from superstition by virtue of the fact that, at least in its more mature and developed forms, it helps people find rationally justified grounds for having faith and hope with respect to matters (e.g. the inevitability of death) that are (i) personally important to them, but (ii) over which they have no control, and (iii) concerning which they also do not have any solid knowledge. My project in this penultimate section of the chapter is to make the ideas just summarized slightly less vague and more intuitively plausible, by saying a few additional things about them.

Cynical people will claim that all attempts to draw a distinction between religion and superstition are hopeless, because it is already clear from ordinary observation that the words "religion" and "superstition" have either the same, or very nearly the same, meaning. In a typical case, for example, a man might pray to God to remove his cancer, not because he has any positive reasons for believing that there really is a God, or that the God whose existence he hypothesizes is willing to do, or even is capable of doing, what he asks, but simply because he cannot prevent himself from doing this.

A magazine article I read years ago gave a humorously satirical expression to the skeptical view we now are discussing. The authors of that article proposed to lampoon the absurd situation that comes into existence whenever certain people (**A**-group) talk about some of their fellow human beings (**B**-group) in either a congratulatory or a condemning way, solely depending upon whether **A**-group thinks that **B**-group is either rich or poor. I am only able to recall three of the examples that appeared in the article. First, the authors claimed that—human nature being what it is at the present time and in this particular society—if you are poor, people say of you that you are weird; if you are rich, they call you eccentric. Next, if you are rich, people describe you as having a fine large family; but if you are poor, they say you breed like a rabbit. Finally, if you are rich, they speak of you as being religious; but if poor, they say you are superstitious. (Again, let me emphasize that the humor here is based on the idea that observers unthinkingly—and from the perspective of wiser reflection, ridiculously—label people, all of whom have more or less the same beliefs and attitudes, and who also engage in more or less the same actions, as either religious or superstitious, simply depending on the observers' impression that the people in question either do or do not have a comparatively large number of financial resources.)

In order to make sense of the third and last mentioned example, let us ask still more precisely what superstition, or a superstition, is. Both humans and non-humans can have, and constantly do have, superstitions. Some psychologists describe a superstition as a mental connection of expectation that an organism forms between two different types of events. Furthermore, from the viewpoint of that organism, it is not possible to justify the legitimacy of that expectation in terms of observable facts. Thus, according to these psychologists, the simplest, clearest, and least misleading example of what it means for a connection of this sort to come into existence, is what happens when one puts a fairly small, uncomplicated creature like a mouse or pigeon into the experimental device known as a Skinner box. For instance, suppose that, just by chance, a pigeon pecks at a blue spot of paint on the otherwise mostly transparent plastic wall of the Skinner box in which it finds itself; and shortly following that peck, a pellet of food comes out of a slot, which the pigeon eats. Even as a result of that one trial, psychologists say, the pigeon already will have formed a weak disposition to peck at the blue spot of paint again, in the unconsciously formed expectation that this action might lead to yet another food pellet. If the pigeon does happen to peck at the spot a second time, and a second pellet does appear, then that disposition will be strengthened. Furthermore, it will continue to be strengthened for an indefinite number of additional, similar trials as well. Nevertheless, this mental connection is only a superstition, because the pigeon has not chosen to form it, and furthermore it does not have any valid reasons to expect to receive a food pellet, after having pecked at the blue spot. (In fact—unknown to this animal—the only "real" connection between the kinds of events that occur in this situation is not a natural one, but one of an "intentional" nature, since it has been brought about through the agency of the desires, will, and planning of the person or persons who set up the experiment in the first place, as a means of getting information about the patterns of thinking and behaving that are typical of certain organisms.)

A clear proof that humans as well as mice and pigeons have an inborn tendency to think and act in a superstitious manner was reported to me by someone who had taken a psychology course at Stanford University in California, which covered the topic (among other things) of Skinnerian, or so-called operative conditioning. Students in this course noticed that the professor had a nervous habit of sometimes scratching the back of his head with the fingers of his right hand while lecturing. They then decided to train him to continue doing that, by systematically giving him the "reward" of looking at him attentively whenever he did it, but looking away, and pretending not to pay attention to what he said, when he did not. My acquaintance reported that, by the time the course ended, the professor had scratched a fairly large bald spot at the back of his head, but still had not become aware of the fact that his students had colluded—made an agreement—to work together to train him to behave in the way that had produced the spot.

Even though superstitions have nothing to do with rational thinking, they still are adaptive in a Darwinian sense of this word, because it usually is a wise policy in practical contexts "not to look a gift horse in the mouth." For example, consider again the story of Ali Baba and the words "Open sesame" that he overheard, from his hiding place in a tree, being spoken by the captain of the forty thieves. Ali Baba had no idea of how or why those words had the power to open the door of the grotto containing the thieves' treasure. (Nor does any hearer or reader of the story have any idea about this point, since it never is explained.) But even so, it would have been foolish for him not to use those words as a means of opening the cave's door, once the thieves had left. In fact, Ali Baba's attitude towards "open sesame" is similar to my own attitude towards pressing the bar on the battery-operated device that automatically opens the door of my garage from a distance. Although I, along with most other users of such devices, have no clear knowledge of how it works, we still are justified in continuing to use it. Thus from my point of view, as opposed, say, to that of the person or people who invented this small machine, the confidence I have in its effectiveness is nothing more than a superstition.

Whether or not it is appropriate to describe someone as thinking and acting in a superstitious way in a given situation depends on certain things about the background of that case. It may be an expression of human nature (relative to present-day society) to assume in a careless and arrogant fashion that the fundamental reason for person **X**'s being poor is the fact that **X** is unintelligent and/or lazy, lacking in ambition, and irresponsible. Objectively considered, however, such an assumption is ridiculous, since reflection on particular cases shows it to be false. For example, quite a few people who are both intelligent and conscientious are happy to work hard at jobs that pay them relatively little money, because of the fact that those professions also provide them with rewards of a different kind than money. For instance, this often is true in the case of scientists, teachers, scholars, religious leaders, policemen, politicians, and military personnel.

Consider again the idea that poor people breed like rabbits. It would be laughably unfair and irrational to say of some teacher, clergyman, or soldier, who had a comparatively large number of children, that he or she bred like a rabbit. This is shown by the fact that, even though he possessed comparatively little wealth, he also would be very likely to take good care of those children, and to give them every available opportunity to develop into happy and productive adults.[12] (Rabbits, we assume, are only interested in copulation, and do not care about the eventual fate of any offspring that copulation might produce.) On the other hand—giving the devil his due—we also must admit that there is another sense of "poverty" that Nietzsche talks about somewhere, which is not a matter of lacking financial assets, but instead is constituted by a particular attitude and style of life. Presumably, for instance, poverty understood in this second way helps explain the sociological fact that most of the people who win a great deal of money in a lottery end up almost penniless very soon after winning. Accordingly, informed and impartial observers ought to agree that a child of a "rich" lottery winner would have a much worse chance of growing into a well adjusted and successful adult, than a child of a "poor" bishop, general, or professor.

Now let us return to the third example mentioned in the article, the one that is relevant to this chapter and book—namely, the supposed lack of contrast between superstition on one side and religion on the other. Is it correct to suppose, as many people do, that these two things are not genuinely different, and therefore a person is entitled to use either the one word or the other in most situations, simply because he or she has a favorable or an unfavorable impression of the people to whom he applies those words? I think the answer is no. Furthermore, one indication of the correctness of this answer is the probable fact that, when our species first came into existence between 200 and 150 thousand years ago, we thought exclusively (along with our Neanderthal neighbors) in terms of animal-like superstitions; but after several tens of thousands of years had passed, we (but perhaps not the Neanderthals) had developed a capacity to engage in more considered and sophisticated thinking as well.

---

[12] As illustrated in my own case, scholarships often help to meet legitimate needs of people of the sort we now are discussing. Since my father was a railroad worker and my mother a homemaker, the only chance I had of going on to study in college and graduate school was provided by financial support supplied by scholarships.

According to the well-known philosopher of religion John Hick (see his book, *God and the Universe of Faiths*), an important historical change took place around the date of 800 B.C., which he describes as "the beginning of The Great Age of Revelation." Hick says this was when so-called natural religion (i.e., a superstitious-like conception of things, based on nothing more than accidentally formed habits of thinking) started to be challenged and then was partially replaced by religious thinking of other types. Hick furthermore proposes to explain this point by claiming that this was when God first began to reveal his true nature and intentions to humans. However, I think Hick's way of talking about this postulated historical event is presumptuous, if only because it presupposes something that neither self-styled atheists and agnostics, nor people who believe in an impersonal rather than a personal god, nor people who believe there are many gods rather than one, would be willing to accept. To be explicit, he assumes (perhaps supported by an uncritical acceptance and relatively literal interpretation of the first few books of the Hebrew Bible) that there exists a single, supremely powerful God who only began to speak to humans in a direct and truthful manner at that particular date in history. But as an alternative to that idea, I suggest it might be more historically accurate as well as less presumptuous to say that what happened at more or less the same date as the one Hick mentions, was that this was when people first began to think about the world and their place in it, in more sophisticated and literal terms than their ancestors had employed, and in ways that also took better account of problems that previously had been left vague and unresolved.[13] To illustrate those problems, consider the contrast (from a still later period) between the Vikings' worship of their war-god Odin, and the Christianity that finally replaced it. In the Viking religion, the reward a man might hope to receive after his death, for an active life of valor and service to the god, was for him to be restored to life once more, and be allowed to dine together with all the other brave warriors in Valhalla. Moreover, after having shared each meal, he again would have the great pleasure of being able to battle to his death in a free-for-all fight with all the other dinner guests.

---

[13] People familiar with my book, *How History Made the Mind* (2003), may remember that I talked there about something roughly similar to this event, which I entitled "the Greek revolution" or "the invention of reason." I claimed that this event first began somewhere on the mainland of Greece around the time of the poet Homer, or in other words, at approximately the date of 1,500 B.C.

Following that again, all of the warriors would be restored to life yet once more, so the same thing could continue to take place, night after night, for an eternity.

In my view, the Viking religion, as compared with Christianity, had the weakness of not setting forth ideals in which all people could share, and in which every person (including women, slaves, cowards, and weaklings, as well as strong and free men) could find a sense of justice, hope, and meaningfulness for their lives. Another failing of this religion is that it did not take the problem seriously of what it might mean to be given life again after death. For example, when, and under what circumstances, could such a thing happen? By what power could it occur? What reasons are there for believing in its reality or at least possibility? Although Christianity did not have any final and obviously correct answers to those same questions, at least it managed to set forth a general framework of thought that allowed one to imagine answers of that sort eventually coming to be clarified, rationalized, and worked out.

Consider again the question, mentioned before, of whether or not a man praying to God to cure his cancer is doing something analogous to (and just as irrational as) what a pigeon does in a Skinner box, because there is no real, causal connection between his prayer and the proposed removal of the cancer, for which he is praying. I believe prayers of this sort need not necessarily be superstitious in all cases. This is because religious people can be motivated to pray as an expression of some kind of reasonable hope, as opposed to their simply being determined to act as they do by automatically formed, superstitious expectations. For example, even if the man knows it is objectively unlikely—i.e. unlikely in terms of the methods, criteria, and standards employed in scientific contexts—(i) that there is a God, and therefore (ii) that God will be willing to grant his wish, it still remains true to say that if there actually is a God (about whom one could not know anything by scientific methods of investigation), then that God both could heal his cancer, and also might actually cure his cancer, if He wished to do so. In other words, the man is in a position to pose the Pascal-like question to himself, of what if anything he has to lose, if he prays for this to happen, and how that loss would compare with the rewards he might receive if (outside the boundaries of possible scientific knowledge) there really was a generous and merciful God who hears and answers prayers.

5.4 Which is more natural and informative: (A) To think about sophisticated human language in terms of recursion and discrete infinity, or (B) To think about such language in terms of psychic distance?

In this last section of the chapter, I want to change the subject slightly. In particular, my project here is to sketch what I believe is a more intuitively natural and, in this sense, an improved way of thinking about sophisticated human language, as compared with the accounts of language of this type accepted by most other philosophers today. Furthermore, this revised view will turn out to be connected with several points already made in preceding pages of this chapter, about the topic of religion.

As noted before, Noam Chomsky claims that the crucial difference between full human language on one side and the communication systems of non-human animals on the other is that each non-human species only has a finite number of signals or calls, but the number of words, phrases, and sentences available to humans like us is potentially unlimited. Chomsky calls this special characteristic of human language "discrete infinity," and claims that it is closely related to the similar notion of recursion—i.e. the unlimited ability to repeat and "nest" linguistic elements of the same general kind, within one another.[14] For example, we already discussed Chomsky's view that the signaling dance of honeybees is more analogous to human language than any other non-human communication system, because—in a manner similar to genuine language—it contains a potentially unlimited number of referring expressions. Nevertheless, he also points out that bee-expressions cannot function in a literally linguistic way, because the special sense of infinity that applies to them is not of the correct type for that purpose. To be more exact, bee expressions are organized on a continuum, analogous to the real numbers, but modern human languages require each of the elements to be related to the others in a way that allows it to be separate, individual, and discrete, analogous to the natural numbers.

I shall not spend time attempting to show that what Chomsky says about these matters is mistaken, because although it seems to me that his ideas of this subject are somewhat awkward, strange, and

[14] In addition to references mentioned earlier, also see Hauser, Chomsky, and Fitch, 2002, where the notion of recursion is emphasized.

counterintuitive, I also am inclined to believe that something like them is true. Rather, my main interest is in asking about the historical background of these Chomskean conceptions. To be more particular, I shall ask the following two questions. (1) How is it reasonable to believe that the connection between recursion/discrete infinity on one side and language on the other first came into existence? And (2) if it also is true to say (as I believe it is) that this connection fails to provide a natural, realistic, or intuitively plausible way of understanding and explaining what modern human language is like, then what accounts for that failing? My answers to these questions will be based on an idea already mentioned at the beginning of this chapter—namely, that what first put humans on the road that eventually would lead them to develop modern language was the invention of religious consciousness. In other words, only after humans had acquired a power to think in quasi-objective terms, did they also come to be in a position to recognize the important role that the notion of discrete infinity would play in the later invention of syntactical language.

As pointed out earlier, ethologists say that some non-human animals like coyotes and crows have a comparatively well-developed ability to count. Nevertheless, none of those creatures has shown itself to be capable of counting in a manner that potentially could continue on without end. The reason this is true, I suggest, is that the kind of counting in which those animals engage is not much more than an extension of their biologically determined, gene-given, "body language." In view of this, the claim I want to make here is that only when humans learned to think in a way that no longer put themselves and their special interests (more particularly, their own bodies) in a central position, did they also acquire a power to overcome the principal limitation that was associated with the thought of other creatures.

I confess I do not know how humans managed to develop the power to count in a potentially inexhaustible fashion. Nevertheless, certain parts of the history of mathematics throw light on another question that is more or less related to this one—namely, the question of how humans came to be able to understand and justify the notion of infinity in theoretical terms. As far as the topic of counting is concerned, it is reasonable to suppose that all the humans who lived before roughly 1000 to 500 B.C. behaved in approximately the same manner as "super-ravens." In other words, those people's ability to order numbers vastly exceeded the limits that apply to other animals; but they still did not have any means of establishing the reality and legitimacy of their conception of infinity, because they had no proof to show that the type of counting they used could—potentially—go on forever, rather than coming to an end at some point short of that goal. The proof that was needed to establish that last claim (i.e. that the natural numbers really did constitute an infinite set) only was supplied at the time and place where the followers of the Greek philosopher and religious leader Pythagoras produced (much against their will) a demonstration of the fact that the diagonal of a square was incommensurate with its side. That demonstration took the form of a *reductio ad absurdum* proof. That is, if one assumes hypothetically that the diagonal *is* commensurate with the side, so that one might express the relation between those two lengths in some complete and exhaustive fashion—e.g. by means of a rational fraction which took one of those lengths as the numerator and the other as the denominator—then one is inevitably led to a contradiction. The contradiction in this case was that one and the same number would have to be both odd and even.[15] Accordingly, then, the negative and indirect proof just mentioned implies that the correct mathematical expression of the relation between the side and the diagonal of a square must include a literally infinitely long number series (e.g.) either of approximating fractions or of approximating decimal points. For instance, if the length of the side of a given square were 1 meter, then its corres-

---

[15] I also have referred to this point in my first authored book, 2003. See the discussion and references on pp.124-7. The statement, explanation, and illustration of the proof offered by the philosopher and mathematician A.W. Moore (also quoted in the previous book) runs as follows:

> Suppose that there *is* a pair of natural numbers such that the square of one is twice the square of the other [which—as we know from the so-called Pythagorean theorem—would have to be the case if a square's side were commensurate with its diagonal]. Then there must be a pair with no common factors (the number 1 does not count as a factor here): for obviously we can, where necessary, divide through. Let p and q be such a pair. Then $p^2=2q^2$. This means that p^2 is even, which means, in turn, that p itself is even. So q must be odd, otherwise 2 would be a common factor. [(Moore's) comment: It is not surprising that the Pythagoreans should have noticed this, given that odd and even occurred in their table of opposites.] But consider: if p is even, then there must be a natural number r such that $p=2r$. Therefore $p^2=4r^2=2q^2$. Thus, $2r^2=q^2$, which means that q^2 is even, which means, in turn, that q is even, contrary to what was proved above. There cannot after all be a pair of numbers such that the square of one is twice the square of the other. (Also see Lloyd 1970. p.35.)

ponding diagonal would be the square root of 2, or in other terms, 1.414213562373 . . . (continuing on for a literal infinity)—meters long.

However, the Chomskean notion of discrete infinity/recursion does not provide the most natural way of understanding what is special about human language; and I believe a better means of doing this (which also fits that notion into a wider context) is to appeal to the quasi-artistic idea of psychic distance. Why is it difficult, and in most cases even impossible, for someone to give an accurate, believable, and aesthetically satisfying representation of himself or herself on the stage or in a movie? The answer is that he is "too close" to his own values, personality, strengths, and weaknesses to be able to examine, know, and depict those things in a correct, sensitive, and understandable way. For example, Alphonse Capone never had the slightest bit of trouble "being himself." But I would bet a great deal of money (if I had a great deal of money) that he could not have done nearly as good a job of presenting a picture of his own personality, attitudes, actions, and career on film, as the actor Robert De Niro did when he portrayed that same personality in the 1987 Hollywood movie, "The Untouchables." For instance, the historical Capone constantly rationalized about, and attempted to justify, everything he did, in terms of such ideas as "the American way," "giving the public what it wants," "what is necessary for running a competitive business," "making friends in high places," "paying back opponents with interest," and so on. Because of that, and because Capone probably had more than half-convinced himself of the truth of most of his own propaganda, he never could gain the kind of perspective on his actions that might have allowed him to see and understand himself as the ruthless, unstable, corrupt, and dangerous criminal most of his fellow Americans (in my view, correctly) considered him to be.

I do not believe that psychic distance is something that is merely supplementary and perhaps also complementary to what Chomsky says about the need for sophisticated language to be organized in terms of discrete infinity. It also is wrong simply to describe this notion as something that allows us to formulate a more familiar, less complicated, and easier to understand version of Chomsky's view. Instead, I suggest that psychic distance provides the kernel of a more insightful account of what language is, and how it works, than Chomsky's theory does, because it presents a more detailed and coherent picture of what the mind must be like in order for the owner of that mind to think in terms of, and by means of, language. In contrast to the opinions of Darwinian theorists like Diamond and

Lieberman, I claim there are two main reasons that language is something that is extremely special and rare on our planet. The first of those reasons is that no organism can succeed in obtaining the use of language, just by employing the automatic and forgetful procedure of "being itself" (Capone style). The second reason is that the activity of being itself is exactly the one in which every member of the overwhelming majority of organisms on Earth constantly engages.[16] To express essentially the same point a different way, nearly every word we utter has the status of being an abstraction of one or another type. And in view of that point, no person can obtain a full or sophisticated language unless he or she develops an ability to think outside the boundaries, terms, and presuppositions of his own personal circumstances, because nothing short of a power of that sort would allow him to represent things abstractly. For these reasons, then, what I claim is that we are justified in thinking of psychic distance as a by-product and a typical expression of the religious principle of separating oneself from the world. Furthermore, this shows what it means to say that the beginning of religious consciousness also was that which first began to steer our ancestors in the direction of developing a power to speak.[17]

In summary, the last point I want to make in this chapter is that I consider one of the three cultural sources of human nature distinguished in this book to be more fundamental than the other two. First, it is implausible to suppose that acquiring a power to domesticate animals was the most basic thing that transformed us into modern humans, because this development was merely a matter of broadening the boundaries of the human situation or niche, in such a way as to allow some non-human animals to occupy a supportive and supplementary position within those same boundaries. Accordingly, it was an expression or symptom of our nature, but was not either its cause or its essence. Similarly, I also do not believe it is correct to say that acquiring language—although this undeniably was a very important step in our history—was the factor that made us into

---

[16] As the professor and poet Mark Van Doren once commented to the author and monk Thomas Merton, "The birds don't know they have names." (See Merton, 1996, p.124.)

[17] If it is correct to say that the members of the Pirahã tribe examined in Chapter 4 "try to avoid all abstractions," then what that means in my view is that those people endeavor to suspend and ignore (as much as possible) a certain invention that some of their and our ancient sapiens ancestors made. The invention to which I refer is that of thinking about things in a detached and quasi-religious manner.

modern humans, as we now understand those words, since language was simply a means of expressing in a more effective, more obvious, and outward fashion, something that already had become more or less present to us inwardly. Again, then, the invention of language was more a symptom of the crucial change than that change itself. Instead, I suggest that acquiring religious consciousness (no matter how deformed, twisted, and misunderstood that step may have been at the time of its beginning) was not just the earliest but also was the most important thing that made us special, which defined our survival niche, and which also laid the foundation for the other developments that came later. The reason this is true is that religious consciousness was what taught us the "trick" of engaging in general and abstract thought, as well as the further trick of separating ourselves mentally from the fast-running experiential stream that dominates, and constantly threatens to swallow up, everything around us.

Chapter 6

Human Nature Conceived as a Lately Discovered, Causally Powerful (but Perilous) Ecological Opportunity

6.1 Darwin compared with Columbus

During the four years I was a student at Kenyon College, located in the small town of Gambier in the middle of Ohio, I occasionally paid a visit to the nearby large city of Columbus. This city had a beautiful central square, one of whose features was an impressive statue of the person after whom the city had been named. I was aware of the fact that the United States observed a yearly holiday in his honor. Furthermore, in spite of his having lived five centuries earlier, I also knew that many other cities both in the United States and around the world had been named for that same person. Thus, whenever I looked at this statue of Christopher Columbus, I found myself wondering if the man it represented really did deserve all the respect and attention that was lavished on him—especially in view of the fact that he had been fundamentally wrong about quite a few things. For example, one mistake Columbus made was that he believed until the day he died that the land he had reached after sailing west from Spain, was India. As a result of that error, we still have the absurd situation today, where many people continue to refer to aboriginal inhabitants of the Americas as "Indians," even though they and everyone else now should be aware of the fact that none of those inhabitants has, or ever did have, anything to do with the country of India.

Defenders of Columbus' reputation will claim that he had a great deal of foresight, determination, intelligence, and courage. They also will remind us of the undeniable point that his voyage turned out to be one of history's sharpest turns. For instance, commentators sometimes give poetic expression to this last idea by saying that, from the moment a sailor in the crow's nest of one of Columbus' three ships first caught sight of land in the west, nothing in the history of our planet was, or ever could be, the same.

However, that sharp historical turn had to be prepared for by a large number of social conditions. For example, those same conditions were not present in the 11th century, when Thorvald the son Eric the

Red led an expedition to colonize a certain part of the North American continent—a place he called Vinland, whose exact location now is disputed.[1] The absence of those conditions in this earlier century is shown by the facts that neither Thorvald's colony nor he himself lasted for a very long time, because of vigorous, armed opposition to the would-be colonizers from hostile native Americans. Thus, for this and other reasons, Thorvald's expedition was not followed by a large number of other Europeans also wanting to travel to America. Still more specifically, (i) by the time of Columbus' voyage, Europe had become overcrowded and oppressive in various respects, so that then quite a few members of its population had formed a keen desire to find other, relatively undeveloped places to which they could migrate. (ii) For quite a few previous centuries, Europeans had wanted to engage in direct trade with rich eastern nations like India and China. But this intention was frustrated by the presence, between them and the other nations in question, of mid-eastern Muslim kingdoms that acted as troublesome and expensive "middle men." (iii) By the late 1400's, Europeans had developed various means by which all of the people encountered in the lands Columbus opened up could be controlled, dominated, and dictated to; but this had not yet happened at the earlier date of Thorvald's voyage. For example, Jared Diamond tells us in his book *Guns, Germs, and Steel* (1999) that, in addition to large, deep-water sailing ships, domesticated war horses, and steel weapons including firearms, one important advantage over people living outside Eurasia that Europeans had developed by the time of the fifteenth- and sixteenth-centuries, was germs of fatal or at least debilitating diseases that had been passed on to them from their farm animals, and which they then were in a position to spread to the relatively less contaminated human inhabitants of the Americas.

Another factor that contributed to Columbus' fame was that most of his mistakes proved to be "forgivable" ones that were relatively easy to comprehend and correct, once people had learned further details. For instance, it later became clear that India and China

---

[1] See Diamond, 2005, pp.205-10. Many people believe this colony was located at L'Anse aux Meadows, Newfoundland and Labrador, Canada, at the end of a long northern peninsula of the island of Newfoundland. (See, for example, Janie Robinson, "The Spirits of the Vikings," *Toronto Star*, May 28, 2009. pp.T1-2.) But Diamond claims this merely was the base camp these Vikings founded in order to have a more or less secure place to stay over the winter, and from which they then proposed to explore more promising locations as soon as summer came.

actually did lie in the direction from Spain that Columbus had claimed; but they were much further away than he had supposed.

I suggest it is a useful exercise to think of Charles Darwin's influence on biology as parallel to the political career of Christopher Columbus. One reason for believing this is that Darwin's publication of *On the Origin of Species* in 1859 turned out to be yet another sharp historical turn. Also—again like Columbus—Darwin's claims steered the people who were sympathetic to him in an essentially correct direction, even though those same individuals later came to see that his theory was partly false, and also largely incomplete, since it failed to take account of many important considerations. To illustrate the point just mentioned, let us return to some of the conclusions Stephen Jay Gould drew about the intellectual legacy of Darwin. (Tim Flannery summarizes quite a few of those conclusions in his 2002.) According to Gould, Darwin should have supplemented his theory with an account of genetics like the one developed by Gregor Mendel, in order to provide an explanation of how characteristics of living creatures arose in the first place, and then were passed along from one generation to another. In addition, Darwin also needed to take better account than he did, of the three topics of "efficacy" (i.e. whether natural selection acts alone, or is supplemented by other, additional evolutionary principles); "scope" (i.e. whether natural selection always operates slowly, inevitably, and gradually, or occasionally works by means of quick, chance events that interrupt its previous course, by setting it on radically new paths); and "agency" (i.e. exactly what the items are on and through which natural selection works). To mention one such failing, Gould believed Darwin had been wrong about efficacy, because he did not pay attention to the fact that creatures' historically determined internal constraints and/ or "spandrels" often helped to determine the course of their evolutionary development. An example mentioned before is whales' inherited bodily form, stemming from the fact that their relatively close ancestors were mammals that walked on four legs, which brought it about that they had to power their swimming with up-and-down rather than side-to-side strokes of their flukes. Similarly, Darwin had almost nothing to say about the topic of "exaptation," or the notion of reusing and recycling body traits originally developed for one purpose, for another purpose instead. (See Gould's 1993c.) Next, Gould believed Darwin had been mistaken about scope, as shown by the fact that we now have overwhelming evidence for the occurrence of occasional catastrophic events, like the asteroid or comet strike off the

Yucatan peninsula 65 million years ago, which at first resulted in mass death, but then was followed by new and in some respects positive evolutionary developments like the appearance of the primates. (Gould 1993b.) Finally, Gould considered Darwin's discussion of agency inadequate, because Darwin assumed that natural selection only acted on and through individual organisms, but most paleontologists now believe new species jump into states that are quite different from, and non-progressively related to, their immediate predecessors ("punctuated equilibria"). Thus, according to Gould, this idea makes it plausible to believe that natural selection can act on species as well as on individuals. (Gould, 1980c.)

Nevertheless, again similar to the case of Columbus, Gould—in spite of the criticisms just mentioned—was happy to describe himself as a follower of Darwin, because he believed Darwin's theory was an essentially correct first step that it was necessary for people to take, to prepare them to participate in what we now think of as modern biology. As one imaginative expression of this point, Gould compared Darwin's account of natural selection to the ancient cathedral or Duomo located in Milan, Italy. He thought this comparison was justified, because this cathedral remained recognizably the same building, and continued to perform essentially the same jobs it always had done, in spite of the fact that, over the centuries, various groups and individuals—e.g. Napoleon—had added many new features, structures, and decorations to it. (On this point, see the discussion and photographs in Gould's 2002, pp.3-6.)

As far as concerns my own work as a philosopher of mind, I also consider myself indebted to Darwin's epic-making work. For example, when I look back over the course of my career, it seems to me that I only managed to acquire a realistic grasp of the nature of human beings and of the characteristic ways humans thought and behaved, once I stopped trying to make sense of those matters from introspective, analytical, psychological, and/or metaphysical viewpoints, and began to view them instead from a perspective inspired by Darwin's scientific "temporalizing" of the Leibnizian Great Chain of Being. I agree with those individuals (see, e.g. David Stamos, 2003) who say that biology, as influenced by Darwin, is the scientific discipline that is most similar to, and most directly useful for, the practice of philosophy considered in general, and of philosophy of mind in particular, because it is the only science that accepts the legitimacy of the question "Why?" Accordingly, then, my Darwin-inspired answer to the question of why (as well as how) humans managed to acquire the nature that now belongs to them is that this nature enabled them to compete successfully with rival species for space, resources, and safety.

However, there is one respect in which it seems to me that a Darwinian view of human beings has fallen short of what it otherwise might have accomplished. My main reason for writing this book was to take a few steps towards trying to heal this crucial weakness in Darwin's legacy. What I am talking about is the fact that Darwin and most of his followers do not pay a proper amount of attention, or attach a proper amount of weight, to the topic of self-created culture. To be more specific, the thesis I defend in this book is that we should take explicit account of the fact that a small number of intellectual inventions made thousands of years ago by a relatively small number of now unknown humans have been important factors that helped create some of the most fundamentally important characteristics that now belong to our species and its nature.

6.2 The ecological concept of Niches is more explanatory than the genealogical notion of species

In this second section, I want to say a few more things about that which Gould called "agency"—i.e. the question of exactly what the items are, upon which natural selection works. Let me begin by distinguishing three different positions people have taken on this subject. There are two opposing extreme views, held respectively by Darwin and by Richard Dawkins, and a third, "reconciling" position that Gould himself supported. As noted before, Darwin believed (as many other theorists continue to do today)[2] that the exclusive objects of natural selection were particular organisms. Thus, even though Darwin entitled his book *On the Origin of Species*, he did not believe there really were such things as species. This is so because, in his view, talk about species was nothing more than a convenient means for allowing people to refer to groups, collections, and populations of individual creatures selected in various arbitrary and culturally determined ways—e.g. by focusing attention on certain properties those organisms happened to share. By contrast, Richard Dawkins says in his book *The Selfish Gene* (1978) that, rather than individual creatures, the real agents of evolution by natural selection are the

[2] See my paper, 1990.

underlying genes that work together first to construct, and then also to control and operate, the organisms that act as their robotic and protective "carriers." Finally, Gould rejected both of the views just mentioned, and maintained instead that evolutionary development took place (sometimes by means of natural selection, sometimes not) on all three of the levels of individual organisms, genes, and species. He said (1994, p.85):

> Natural selection is not fully sufficient to explain evolutionary change . . . [because] many other causes are powerful, particularly at levels of biological organization both above and below the traditional Darwinian focus on organisms and their struggles for reproductive success. At the lowest level of substitution in individual base pairs of DNA, change is often effectively neutral and therefore random. At higher levels, involving entire species or faunas, punctuated equilibrium can produce evolutionary trends by selection of species based on their rates of origin and extirpation, whereas mass extinctions wipe out substantial parts of biotas for reasons unrelated to adaptive struggles of constituent species in "normal" times between such events.

Of these alternatives, I am attracted to the "combination" position defended by Gould, because I believe the world we live in is sufficiently large and complex to be able to accommodate it. For example, this fairly complex way of looking at things is repeatedly instantiated in many of the situations we encounter in ordinary life. To consider a very trivial instance: Sophisticated dance instructors tell their male students that any of them who proposes to learn to perform the Argentine Tango in a sensitive and competent manner, should constantly keep in mind that he is dancing at the same time with three different but complementary "partners"—(i) a living breathing female, (ii) the music, and (iii) the floor.

What arguments did Gould give in defense of his three-part conception of agency? For one thing, Darwin claimed that nearly all species changed in a gradual and constant manner over the course of time, and therefore it was inevitable that people should disagree about the exact boundaries of a species, since different observing cultural groups, with their various preoccupations and needs, and perhaps even the different times at which they made their observations, would be bound to notice and emphasize different aspects and phases of the same organisms. But Gould pointed out that invertebrate paleontologists like himself and Niles Eldredge were lucky enough to work with nearly complete and unbroken fossil records. And it was clear from those records that many or even most species did not change gradually from one to another. Instead, each of the members of a long chain of related species went through a pattern of coming into existence, then staying the same for a relatively long period, and then finally going out of existence, in a series of "jumps." Gould argued further that, if a Darwin-like view of these matters was correct, then human observers ought to disagree much more often, and to a much greater extent, than they actually did, about what did and did not count as a given species. As a matter of fact, however, reports of anthropologists show that there is a remarkable amount of agreement among people who belong to different cultural groups, about the properties and boundaries that belong to each of the species of creatures that all of them observe. For example, even though the tribes of New Guinea live close together on a fairly small island, most of their languages are quite dissimilar; and there are many other, sharply defined cultural differences between those tribes as well. But in spite of those differences, and furthermore in spite of the great cultural gulf that separates New Guineans on one side from visiting Australians, Americans, Japanese, and Europeans on the other, New Guineans distinguish virtually the same number of bird species that are found in their local areas, and describe those species in virtually the same ways as both their neighbors do, and as bird experts from Europe, America, Japan, and Australia also recognize. (There is a summary of this and related points in Stephen Jay Gould's essay, "A Quahog Is a Quahog," 1980d.)

Nevertheless, with all the preceding having been said, I suggest that "complementarity" accounts of agency similar to that which Gould proposed suffer from at least one important problem. The problem is this: Although people commonly assume that genes and species are real entities that empirical scientists are able to investigate in a fruitful fashion, our usual way of thinking and talking about them implies that they are nothing more than abstractions. Yet if it really were true that they are abstractions, then it ought to be impossible for them to have the kind of causal force that would allow them to function as agents of natural selection. For example, a tiger can bite you; but neither the word "tiger" nor the concept of *tiger* is capable of doing that. Similarly, it also seems that neither the species of tigers, nor any of the long-lasting genes that presumably belong to and help to define tigers, can bring about any physical changes in the world, in view of the fact that those items are nothing more than abstract entities (universals), like the idea of freedom, or the game of poker, or the definition of the word "hippogriff," rather than concrete particulars.[3]

Let me elaborate the preceding point a little further. People often speak in a casual manner about genes and/or species "controlling" the properties and development of animals. But it should be clear on reflection that this way of talking is nothing more than a metaphor. For instance, Darwin thought of species as genealogical relations of descent in which certain organisms stood to each other. But genealogical relations (like all other abstract relations, like the relation of "below" or "on top of" or "older than") do not have any power to influence, change, and arrange things in the physical environment in such a way as to direct, order, and control the things that stand in those relations. Thus, it presumably follows that species also cannot be items through which natural selection is able to act. For example, neither genes nor species have hands, legs, bodies, brains, or physical strength. They also are incapable of making choices or of putting choices into effect. In fact, if species are nothing more than abstract entities then, as any student of philosophy who is informed about the history of his or her subject will be glad to point out, they cannot even occupy any definite positions in time and space. For this reason, I do not believe Gould was justified in proposing to accept the view that, in addition to particular organisms, species and genes also can act as both objects and agents of natural selection.

It seems to me that the most serious attempt yet made to deal with the problem of whether species can be *bona fide* agents of natural selection, is associated with an imaginative picture of evolution's workings proposed by the American geneticist Sewall Wright. I now propose to talk about a short passage in which Ian Tattersall (1999, pp.22-3) discusses the meaning and background of Wright's idea. Tattersall says the so-called "evolutionary synthesis" in biology, which combined the notion of natural selection on one side, with shifts in gene frequencies within populations on the other, managed to satisfy both naturalist biologists whose main concern was with the diversity of living nature, and other biologists whose focus was on the field of genetics. Furthermore, he said, one of the most compelling

---

[3] As George Williams says (quoted by Daniel Dennett in 2006, p.439), "A gene is not a DNA molecule; it is the transcribable information coded by the molecule." Again, as Michael Ghiselin claims (quoted by David Stamos in 2003, p.1), "The species problem has to do with biology, but it is fundamentally a philosophical problem—a matter for the 'theory of universals.'"

images that tied those two themes together was proposed by the American geneticist Sewall Wright, who introduced idea of an "adaptive landscape." Tattersall's way of explaining this idea was to claim that each individual in a population possessed a unique combination of genes (genotype), and some of those combinations produced individuals that were more fit than others to survive and reproduce in a given environment. Wright then proposed the equivalent of a topographic map in which the contour lines did not connect points of equal elevation, but points of equal fitness (i.e., the ability to survive and reproduce). Accordingly, this implied that relatively fitter genotypes would gravitate to and be found on the hilltops of the adaptive landscape, while the less fit genotypes would occupy the valleys. At this point in the passage, Tattersall made the following remark:

> As Wright saw it, the problem faced by each species was to maximize the number of individuals on the hilltops and to reduce as much as possible the number of those in the valleys, a result that should inevitably emerge from the operation of natural selection.

To illustrate the difficulty about which I was speaking before, let me now pause to ask a rhetorical question about the phrase just quoted. The question is this: Since a species—or a gene—is not something of the same kind as a particular organism, what sense can it make for either Wright or Tattersall to talk about a species being "faced by" a problem, or of its employing various means to "overcome" that problem?

I now continue my paraphrase of Tattersall's passage. He said that Wright's metaphor, which linked natural selection in a neat way to the distributions of genotypes, proved to be extremely influential. Several important evolutionary thinkers adopted it and elaborated it in various respects, in an attempt to show how natural selection, over the long run, could produce not just change in lineages but also could account for the various discontinuities that are found in nature as well. Darwin admitted that each species of complex organisms was a separate genetic package that was not able to exchange genes with any other species, no matter how closely related those two species might be. But biologists eventually came to interpret the peaks in Wright's landscape as ecological niches, to which the species that occupied them were well adapted. Species would cling to those peaks (niches), and avoid the hostile territory of the valleys. However, the landscape itself was not static, but, in the words of the paleontologist George Gaylord Simpson, was "more like a choppy sea." Therefore natural selection constantly had to work at the task of keeping species balanced on the peaks that constantly shifted beneath them as environments changed. For example, sometimes a peak would split in two, carrying different populations of the same species in different directions. Nevertheless, Tattersall said, in view of the fact that selection for different conditions constantly operated on each new peak, the emergence of new species seemed to be something that inevitably was bound to happen.

To summarize, Wright's metaphor inspired an intellectual change. Recent biologists shifted from talking about species as possible objects of natural selection, to speaking instead—in more precise and focused terms—about this being true with respect to ecological niches. I want to point out here that one reason for supposing this change really did bring about an increase in theoretical coherence is that it allowed and entitled one to suppose that niches have a source of causal power that was not available to species. The causal power to which I refer is causality of a pull rather than push type. Still more explicitly, ecological niches can play an active role in the various processes of natural selection, because each one of them that happens to be currently active and operating at the present time, and therefore possesses an attractive power that physically draws (sucks) living things into it. This, I suggest, is the means by which a niche is able to influence and physically determine many properties of the organisms that happen to occupy that niche.

The scientific usefulness of niches is a source of evidence in favor of the truth of something like the Platonic theory of universals. That is, more or less in the style of Plato, working biologists can discover which species of (or ecological niches for) living creatures actually do exist, and which other, imaginary, hypothetical, or merely possible species (or niches) do not exist. However, unlike Plato, biologists do not make decisions about these matters by appealing to a priori powers like a "God-given inner light." Instead, it is their regular practice to decide which niches and species are real, on the basis of careful observations of, and empirically justified inferences that they draw concerning, the same ordinary, physically existing world in which all of us, and all of the living creatures that we study and observe, live, move and have our being. (I also have discussed this same point in Chapter 4 of 2003.)

At the risk of boring my readers, let me now say much the same thing in a slightly different set of terms. If it were true to claim that species were non-physically existing, merely abstract entities, then Darwin would have been right to maintain that there are no such things as species (and *a fortiori* that species cannot be either agents or objects of natural selection). Nevertheless, I believe there both can be, and are, generals that exist in the physical and observable world around us. Moreover, some of those generals play the role of expressing and representing certain species. The existing generals, about which, I now am speaking, are the ones that modern biologists refer to, and propose to identify as, currently active and effective ecological niches.

Let me now conclude this section with a quick historical overview. In the last part of the thirteenth century and the first part of the fourteenth century, the scientifically motivated English philosopher, William of Ockham, popularized the doctrine called "nominalism." This was the idea that there were no generals or universals existing in nature, as proved by the fact that it was wrong to multiply entities beyond necessity, and it was not necessary to posit the existence of universals, since entities of that type did not have the power to explain anything. The comment I want to make about this is that even if people once had the impression that this argument of Ockham's correctly applied to science, as they thought about science in the fourteenth century, it does not also apply to the sort of modern, post-Darwinian biology with which we have become familiar in the twenty-first century. Thus, consider a pair of instances I also discussed in my 2003 book. First, what accounts for the fact that kangaroo crossing signs appear on Australian highways in roughly the same kinds of places that deer crossing signs also appear on North American highways? The short answer is that, in spite of the observed fact that kangaroos hop on two legs while deer run on four, those very distantly related species of creatures are occupants of the same ecological niche. The niche in question (pull)-caused those two species of animals both to seek out, and to live in, relevantly similar environments, in spite of the fact that the species were not closely related.[4] Second, what explains the historical fact that the so-called

---

[4] One interesting point about this example is that it shows the kinds of characteristics that niches are able to determine causally, as opposed to the kinds they do not have the power to bring into existence. That is, the niche that deer and kangaroos share determines that any animal that occupies that niche must be a large, swift moving

"bear-dog" became extinct soon after the joining of the continents of North and South America? The species of animals to which biologists gave this nickname was a top marsupial predator that originated in South America. But as soon as drifting tectonic plates brought these two continents together, it proved to be the case that the bear-dog was not able to run sufficiently fast or sufficiently long to compete successfully with the North American placental-mammal dogs (wolves, coyotes, and foxes) that now occupied the same territory. Similarly, it also turned out that bear-dogs were not large and powerful enough to compete with the North American placental bears that now were bound to challenge them. In other words, this species became extinct because the "mixed" predator ecological niche it once had occupied had ceased to exist (i.e., had stopped being active and effective) once the two continents came together. Instead, the no longer useful niche just mentioned had been replaced by two other, relatively more specialized niches.

6.3 Considered together, the three cultural inventions discussed in this book add up to our ancestors' discovery of an unoccupied Niche, physically present in the natural world

Most species come into existence by following relatively predictable paths that other, similar species have taken before them. For example it is nearly a sure bet, in view of the fact that all present-day African antelopes live and survive in essentially the same manner, that any new species of antelopes that develops in that same environment will turn out to be similar to the ones that already are found there. Occasionally, however, a group of organisms manages to find and to begin to employ a survival trick that is vacant, because it has not previously been discovered by creatures of any other sort. In that case, the group in question becomes a new species (in fact, sometimes even a whole new genus or family of species) that is dramatically different in various respects from all other species. I began this book with a discussion of hummingbirds, because those birds seem to be products of just such an exceptional situation. Once those animals started to

herbivore that is capable of eating grass; but it does not causally determine the exact color or shape the animal who occupies the niche must have, or what precise method of reproduction it must employ, or the number of legs on which it has to move.

follow the apparently unbird-like practice of feeding from the nectar of flowers in a way similar to bees, then natural selection, working through an ecological niche, caused their bodies to change so as to allow them to take better advantage of that style of life. For example, their bills became longer, thinner, and in some cases curved; the colors of their feathers became brighter and iridescent; they became smaller, lighter, faster, more agile, and more energetic. As mentioned before, it seems to me that people like us ought to find birds of that type interesting, because our species probably took a path that was parallel to theirs.

Another, perhaps clearer example of what I am talking about is the marine family called exocoetidae or flying fish. These are fish of small to medium size, comprised of about 50 species grouped into 7 to 9 genera, which are found in all the major oceans of the world, especially in the warm tropical and subtropical waters of the Atlantic, Pacific, and Indian oceans. The most striking features of these fish are their unusually large pectoral fins and an enlarged lower tail lobe, which allow them—when startled—to take short gliding flights through the air above the surface of the water, which enables them in turn to escape from larger predators like tuna and swordfish.[5] Thus it is possible to account for their distinctive bodily form in Darwinian terms. That is, if a group of fish that lived a long time ago happened to develop a larger set of pectoral fins than those belonging to other members of their species, which allowed them to glide some distance over the water when they were threatened by predators, then they would be more likely to escape being eaten, and also would be more likely to pass on their genes to subsequent generations. This eventually led that group of fish to develop into a separate species of their own, which then started to compete with the original species from which they had arisen, and possibly either did in the past or would in

---

[5] This trick flying fish employ usually is effective. But in some circumstances it no longer works. For example, consider the following passage from Hemingway's novel, *The Old Man and the Sea* (p.34):

> As he watched the bird dipped again slanting his wings for the dive and then swinging them wildly and ineffectually as he followed the flying fish. The old man could see the slight bulge in the water that the big dolphin raised as they followed the escaping fish. The dolphin were cutting through the water below the flight of the fish and would be in the water, driving at speed, when the fish dropped. It is a big school of dolphin, he thought. They are widespread and the flying fish have little chance.

the future, drive that original species into extinction. In other words, that which brought flying fish into existence was an ecological niche that some of their ancestors located then occupied.

Why do I think something similar also happened in the case of our species? The reason is that we know from the historical record that about 100,000 years ago some members of our species became different from all the other hominids, with whom they previously had shared essentially the same behaviors and forms of life, in something like the way that hummingbirds became distinct from wrens, and flying fish became separate from their non-flying progenitors. The point I have been emphasizing in this book is that, in the human case, this transformation was not just a biological one, but was partly cultural. Biologically speaking, we are and remain (as Jared Diamond says, see his 1992/2006, p.1) "a species of big mammal, down to the minutest details of our anatomy and our molecules." Nevertheless, we did not become different from close relatives like the Neanderthals just by changing our physical characteristics and behavior. Instead, the key causes of the change just mentioned were factors like the invention of a language that was more powerful, organized, flexible, and precise than any that might have been available to other hominids, and the creation of societies that were larger, more inventive, and more specialized in terms of labor and other roles, than was the case with other hominid social groups.

In some respects, our species might have been more "suited" to make a transition of this special type than other hominid species. This is because the members of our species are now (and presumably were for a long time in the past) much physically weaker than any other of the large primates, as a result of the fact that a large proportion of our energy and blood flow is directed into brain cells rather than muscles. For example, Sue Savage-Rumbaugh says (1998, p.8) that Kanzi, the adult, 165 pound bonobo or Pygmy Chimpanzee she has observed over a long period, and on whom she has conducted an extensive series of experiments, is "very strong—five times stronger than a 165-pound human male in excellent physical shape." (I vaguely remember reading a newspaper article once about a fair that took place somewhere in Tennessee, where a boxing contest was organized between any person who cared to participate, and a chimpanzee fitted with boxing gloves. Although I do not recall many details about the article, I do remember that not a single one of the human participants succeeded in making good on any of the boastful threats he initially might have made about what he intended to do to the chimp.) According to Savage Rumbaugh, one of the anatomical and physiological reasons for the greater strength of our primate relations is that the muscles of chimpanzees, gorillas, orangutans, etc. are denser than ours, since they contain a relatively greater number of fibers than ours do. (See ibid., p.31.) Similarly, archeologists say the Neanderthal skeletons they find are composed of bones that are larger and more compact than ours, which is consistent with the idea that those now extinct people were more robust and physically more powerful than we are now and perhaps also as we were at the time when we lived in close association with the Neanderthals. (On this point, see Hall, 2008.) Accordingly, if our ancestors really did exterminate their Neanderthal neighbors, they must have done this by means of tricks and stratagems like ambushes, traps, and the use of fire, rather than by engaging Neanderthals in close physical combat. Again, as already said, it might have been comparatively easy for members of our branch of the human family to become specialized in intellectual matters, as contrasted with feats of physical strength, because the bodies with which they were born already devoted a greater amount of their resources and energy to activities of the first type rather than the second.[6]

In this book, I have been trying to specify the centrally important parts of the ecological niche that enabled our species to distinguish itself from all other known creatures, including all other species of humans. Since I am not a professional archeologist, I have not tried to specify precise dates at which our ancestors began to occupy each of the three main aspects of what I think of as our species' special niche. Nevertheless, my impression is that those intellectual inventions probably appeared, fairly close together in time, starting at the date of roughly 100,000 years ago. Furthermore, as opposed to the way many other theorists think about these matters, I have not argued that language was the most crucial or fundamental part of our niche, or even that it was the earliest factor that played a role in it. Instead, I believe a relatively objective type of thinking ("religious consciousness") probably appeared before we went on to acquire the ability to domesticate other species, and before we acquired a complex language, and that consciousness of that sort then acted as a prerequisite for those other two developments to take place. In summary, then,

---

[6] Consider the following remark attributed to William Hazlitt (1778-1830) "Cunning . . . is the sense of our weakness, and an attempt to effect by concealment what we cannot do openly and by force."

my thesis is that when these three mutually supporting inventions began to work together for the first time, a style of life became open to our ancestors that was vastly different from anything known before by primates considered in general, or by hominids in particular. In fact, as I said before, the change just mentioned was nothing less than a quantum leap beyond the intellectual capacities that belonged to any other creatures on our planet.

Finally, the last comment I want to make in this section is that the special sort of intelligence developed by members of our species was not the only kind of intelligence that either is possible, or even that is actual and existent in our world. For example, there recently appeared a short note in my newspaper ("Brains in the Deep," *The Globe and Mail*, Tuesday, July 27, 2010, p.L6), on the subject of current, research-based thinking about octopuses (largely quoted from another article written by Emily Anthes for *The Boston Globe*). According to this note, scientists' discoveries about octopuses have called into question many of our usually accepted ideas about intelligence. In particular, we have tended to suppose for a long time that intelligence only could have evolved in creatures who lived in groups, as something that helped them navigate the complexities that were implicit in their social environment. But octopuses, which are almost as far removed from primates like us as it is possible for any animals to get, lead a solitary life. They are a kind of mollusk, like clams. But unlike clams, they have no shell to protect them against predation. Thus, it probably is true to say that the considerable amount of intelligence these animals now display was an indirect result of Darwinian processes whereby many successive generations of single individuals—using any and all the ruses and tricks they could devise—struggled to avoid being eaten.

6.4 Historically accumulated layers of human nature, and the contrast between good and bad ways of combining those layers

> Never think that war, no matter how necessary, nor how justified, is not a crime.
>
> Ernest Hemingway

> Nothing ever really sets human nature free, but self-control.
>
> Phyllis Bottome

In his novel, *War and Peace,* one of Leo Tolstoy's main characters describes war as "the vilest thing human beings ever do." That which made war disgusting, in his (and perhaps also Tolstoy's) view, was the fact that it was unnatural, twisted, and perverted, in the sense of being fundamentally opposed to both human reason and human nature. However, in other places in the same book, Tolstoy said things that seemed to be at odds with this idea. For example, one comment he made about the French invasion of his country in the early nineteenth century was that although Napoleon claimed to be leading a million men into Russia, he actually was following them. Furthermore, according to Tolstoy, the real cause that either enticed or forced this vast number of people to go where they went and do what they did, was exactly the same thing as that which made birds in the northern hemisphere fly south in the fall. This last statement apparently implied that war could not really be opposed to human nature after all; and therefore one is forced to conclude instead (in the style of Sigmund Freud) that war is intimately included in that same nature, as part of its "darker side." Another, slightly different way of making the same point is to say that his remarks about Napoleon's fraudulent claims of leadership imply that what is vile, depraved, and disgusting about war is not war itself, but the minds and souls of all those humans who agree (or allow themselves to be "forced") to participate in it.

My general project in preceding chapters has been to name, describe, and interrelate three ancient inventions that collectively constitute what we now think of as our special ecological niche and our special human nature. But even on the unlikely assumption that everything I have said about this matter is correct, all this talk could not amount to more than an elementary introduction to a vastly complicated subject, since there are many other problems about our nature that still remain to be solved. For example, why does our nature lead to bad and repulsive things like war, poverty, malicious cruelty, and pollution, as well as good things like warm houses, philosophy books, ballet performances, and water treatment plants? Also, what hope do humans like us have of eventually becoming able to rethink, discipline, and reform this nature so as to eliminate the bad things connected with it, or at least to keep those things under control? To express the same point metaphorically, what I have done so far in this book is to slice the cheese in one direction (say, vertically); but I now am proposing to end the book by reminding readers of other possible ways in which it might be sliced instead (say, horizontally), in order to present a more balanced and complete snapshot of what we are like.

I do not pretend to know exactly what Tolstoy had in mind when he talked about war in his novel. Nevertheless, one possible way of resolving the apparent contradiction just mentioned is to make some distinctions. First, when people speak of "human nature," they sometimes refer to the entire package of tendencies (desires, fears, strengths, weaknesses, abilities, disabilities, etc.) that present-day people constantly carry around with them. But they also sometimes use this phrase to refer only to the particular part of the package, which our species and culture have added to that which already belonged to our pre-modern ancestors. Thus, Tolstoy might have considered war contrary to human nature because he believed that certain primitive tendencies included in the (broad) nature that we had inherited from our forebears was what repeatedly led us to engage in war; but war was contrary to our more distinctive, more recently developed and invented, (narrow) human nature.

Second, it also is possible to distinguish between goals, values, and personal commitments to action, on one hand, and the practical means someone might employ to accomplish one or more of those goals, on the other hand. Each of the principal, historically determined levels of our whole nature—(say) the reptilian level, the mammal level, the primate level, the hominid level—probably is connected both with certain goals, and also with various possible means of realizing those goals. Our relatively unsophisticated early ancestors, before the date of about one hundred thousand years ago, probably occasionally employed their newly developed or invented powers of thinking to accomplish certain "old" goals and projects that were typical expressions of their older, primate nature. For example, this may have happened in a case mentioned in Chapter 5, where our ancestors decided to target, and finally to exterminate, their cousin hominid species, the Neanderthals. But Tolstoy considered it inexcusably perverse for people to continue doing something of that same sort today, at a time when their faculties of thinking, choosing, and acting had become more powerful, clear, and self-aware than they had been previously. In other words, the reason Tolstoy considered war repulsive and vile is that he believed it was a result of people putting their God-given abilities to think in objective, considered, and moral ways, at the service of primitive, animal-like desires and goals, in spite of their having developed a potentiality to be something more and better than animals.

Contrary to what romantic people like to suppose about the "natural nobility" of all non-human creatures, we know from observation that there are many expressions of corporate, planned, war-like, and even genocidal behavior among our closest living relatives, the Common Chimpanzees. (See Diamond 1992/2006, pp.290-4.) However, in spite of these animals' strength—and luckily for the Tennessee farm boys who got into a boxing ring with one of them—their killing behavior is not very efficient. For example, wild chimpanzees often form gangs to hunt, attack, and kill particular chimpanzees of other troupes. But they have not yet learned to bring their attack to a quick end, by means (e.g.) of strangulation. Similarly, they also have not learned to set traps for their enemies or to organize ambushes. Thus, only when chimp-like behavior and values are combined with a modern human's ability to think and reason does the murderously effective institution we now call war come into being. War in our era is typically conducted in a calculating and cold-blooded manner that is unknown among other animals. To mention just one case, arbitrarily selected from many millions that might have been chosen instead, consider the pleasure some Saracen warriors took in skinning captured European Crusaders alive, then arranging to have their neatly folded skins sent back to their wives in France, Italy, and Germany, as beautifully wrapped "gifts." In summary, one might say that what created the perversion of war is the lethal combination of ape desires and goals combined with modern human means of putting those goals into effect.

What I have said in this book is only a small part of all that needs to be said about human nature. Nevertheless, I still believe the basic account for which I have argued is correct. It is possible to sum up that account in the following five statements: (1) Human nature has not remained the same throughout the lifetime of our species; and therefore one cannot understand it correctly unless one takes account of its historical development during that period. (2) Culture, in addition to biology, has played a role in making human nature what it is now. (3) Rather than something that appeared all at once, this nature was a result of the combining of several separate but related intellectual innovations that happened at different times. (4) Human nature was not brought into existence by our species alone, but also took shape and benefited from contributions made by non-human organisms. (5) The nature we have at the present time has many substantial and intimate relations to the ideas of reason, morality, objective thinking, and faith.

Finally, let me explain something that might have struck some readers as a contradiction. In my 2003 book, *How History Made the Mind,* I discussed the so-called "pre-Classical Greek invention of Reason" in which people learned—at about the date of 1500 B.C.—to leave behind their former reliance on magical and mythical habits of thought (characteristic of the ancient Egyptians), and adopt instead a new ideal and standard of thinking in ways that were rational, literal, and based on evidence that could be assessed and accepted by every person alike. The sub-title of that book was: *The Cultural Origins of Objective Thinking*. But now, in Chapter 5 of this second authored book, I also have described the pre-historical invention of religious thought—which presumably happened at the much earlier date of 100,000 years ago—as the advent of "objective consciousness." Which is it to be? Do I believe that objective thought came into existence among human beings during pre-historical times, at about 100,000 years ago? Or do I think it made its first appearance in historical times, roughly 5,000 years ago? My answer is that it is correct and fair to describe the entire historical process in which our ancestors gradually and painfully fashioned present-day human nature, which I have recounted in both of these books considered together, as the development of thinking which is rational and objective. However, the beginning of this process that happened 100,000 years ago with the invention of religious consciousness was a much less perfect, consistent, and complete form of objective thinking, than the sort of objective thinking that pre-Classical mainland Greeks inaugurated 5,000 years earlier than today. Thus, I claim there is no contradiction here, but only a story about how an extremely rough, shifting, and inchoate form of rational thought finally became replaced by the more robust, consistent, and substantial form of this same general kind of thought that (most) people (mostly) employ today.

Bibliography

Anonymous. 2006. "Mondo hummingbirds," *The Globe and Mail,* Monday, March 20, p.A14.

Anonymous. 2007. "Homo unpleasantus?" *The Globe and Mail,* Thursday, February 15, p.L6.

Anonymous. 2008. "The Science of Religion: Where Angels no longer Fear to Tread," *The Economist,* March 22, pp.89-92.

Anonymous. 2008. "The First Farmers," *The Globe and Mail,* Monday, April 7, p.L6.

Anonymous. 2010. "Brains in the Deep," *The Globe and Mail,* Tuesday, July 27, p.L6

Anonymous. 2010. "Who's for Lunch?" *The Globe and Mail,* Monday, August 30, p.L6.

Ardrey, Robert. 1972. *African Genesis,* Dell Publishing, New York, NY.

Arsuaga, Juan Luis. 2002. *The Neanderthal's Necklace: In Search of the First Thinkers,* Four Walls Eight Windows, New York, NY.

Attenborough, David. 1979. *Life on Earth: A Natural History,* Little, Brown, and Company, Boston and Toronto.

——. 1984. *The Living Planet: A Portrait of the Earth,* Collins, London, U.K.

Augustine (St.), see Bourke, 1978.

Bickerton, Derek. 1990. *Language and Species,* The University of Chicago Press, Chicago, IL.

Birx, H. James. 1998. "Introduction," pp.ix-xxviii in Darwin, 1871/1998.

Bourke, Vernon J. (ed). 1978. *The Essential Augustine,* Hackett, Indianapolis, IN.

Burkert, Walter. 1996. *Creation of the Sacred: Tracks of Biology in Early Religion,* Harvard University Press, Cambridge, MA and London, U.K.

Calvin, William H. 2002. *A Brain for All Seasons: Human Evolution and Abrupt Climate Change,* University of Chicago Press, Chicago, IL.

Chomsky, Noam. 1988. *Language and the Problems of Knowledge: the Managua Lectures,* MIT Press,Cambridge, MA.

——. 1997a. "Language and Cognition," in Johnson and Erneling, 1997, pp.15-31.

——. 1997b. "Language from an Internalist Perspective," in Johnson and Erneling, 1997, pp.118-35.

——. 2000. *The Architecture of Language,* Oxford University Press, New Deli.

Churchill, Winston S. 1950. *The Hinge of Fate,* Houghton Mifflin, Boston, MA.

Constable, George and the editors of Time-Life Books. 1973. *The Neanderthals,* Time-Life Books, New York, NY.

Darwin, Charles. 1859/2000. *On the Origin of Species by Means of Natural Selection, or the Preservation of Favoured Races in the Struggle for Life* (first edition), edited by Ernst Mayr, Harvard University Press, Cambridge, MA.

——. 1871/1998. *The Descent of Man; and Selection in Relation to Sex* (with an Introduction by H. James Birx), Prometheus Books, Amherst, NY.

Dawkins, Richard. 1978. *The Selfish Gene,* Granada Publishing, London, U.K.

Deacon, Terrence W. 1997. *The Symbolic Species: The Co-evolution of Language and the Brain,* W.W. Norton, New York, NY.

Dennett, Daniel C. 1995. *Darwin's Dangerous Idea: Evolution and the Meaning of Life,* Simon and Schuster, New York, NY.

——. 2006. *Breaking the Spell: Religion as a Natural Phenomenon,* Penguin, New York, NY.

Diamond, Jared. 1992/2006. *The Third Chimpanzee: The Evolution and Future of the Human Animal,* HarperCollins, New York, NY.

——. 1999. *Guns, Germs, and Steel: The Fates of Human Societies,* W.W. Norton, New York, NY.

——. 2005. *Collapse: How Societies Choose to Fail or Succeed,* Penguin Books, New York, NY.

Donald, Merlin. 1991. *The Evolution of the Modern Mind: Three Stages in the Evolution of Culture and Cognition,* Harvard University Press, Cambridge, MA.

Dube, Rebecca. 2007. "A Murder of Crows," *The Globe and Mail,* Friday, December 14, pp.L1-2.

Editors of the How and Why Library. 1964-76. Childcraft, the How and Why Library, Volume 5: About Animals, World Book-Childcraft International, Inc.

Eldredge, N. and S.J. Gould. 1972. "Punctuated Equilibria: An Alternative to Phyletic Gradualism," pp.82-115 in Schopf, 1972.

Erneling, Christina E. and David Martel Johnson (eds.). 2005. *The Mind as a Scientific Object: Between Brain and Culture,* Oxford University Press, New York, NY.

Feynman, Richard P. 1995. *Six Easy Pieces: Essentials of Physics Explained by its Most Brilliant Teacher,* Addison-Wesley, Reading, MA.

Fischer, Steven Roger. 1999. *A History of Language,* Reaktion Books Ltd., London, U.K.

Flannery, Tim. 2002. "A New Darwinism?" *The New York Review of Books,* May 23, pp.52-4.

Gordon, Peter. 2004. "Numerical Cognition without Words: Evidence from Amazonia," *Science,* Vol.306 (October 15), pp.496-9.

Gould, Steven Jay. 1980a. *The Panda's Thumb: More Reflections in Natural History,* W.W. Norton, New York, NY.

——. 1980b. "Natural Selection and the Human Brain: Darwin vs. Wallace," pp.47-58 in 1980a.

——. 1980c. "The Episodic Nature of Evolutionary Change," pp.179-85 in 1980a.

——. 1980d. "A Quahog Is a Quahog," pp.204-13 in 1980a.

——. 1993a. *Eight Little Piggies: Reflections in Natural History,*W.W. Norton, New York, NY.

——. 1993b. "The Wheel of Fortune and the Wedge of Progress," pp.300-12 in 1993a.

——. 1994. "The Evolution of Life on Earth," *Scientific American,* October, pp.85-91.

——. 1997a. "Darwinian Fundamentalism," (Part 1), *The New York Review of Books,* June 12, pp.34-7.

____. 1997b. "Evolution: The Pleasures of Pluralism," (Part 2), *The New York Review of Books,* June 26, pp.47-52.

——. 2002. *The Structure of Evolutionary Theory,* Harvard University Press, Cambridge MA.

Gould, Steven Jay and E. S. Vrba. 1982. "Exaptation—a Missing term in the Science of Form," *Paleobiology,* 8(1), pp.4-15.

Hauser, Marc D. 2000. *Wild Minds: What Animals Really Think,* Henry Holt and Company, New York, N.Y.

Hauser, Marc D., Noam Chomsky, and W. Tecumseh Fitch. 2002. "The Faculty of Language: What is it, Who has it, and How did it Evolve?" *Science,* Vol. 298, November 22, pp.1569-79.

Hemingway, Ernest. 1952/2003. *The Old Man and the Sea,* Scribner, New York, NY.

Hick, John. 1973a. *God and the Universe of Faiths,* Macmillan, London.

Hurford, J.R. 1987. *Language and Number,* Blackwell, Oxford, U.K.

Hurst, Lynda. 2008. "Sprinting down the Evolutionary Highway," *Toronto Star,* Sunday, February 3, p.ID4.

Ingram, Jay. 2004. "Tiny Tribe has Never played the Numbers Game," *Toronto Star,* Sunday, August 29, 2004, p.A14.

James, William. 1902/2000. *The Varieties of Religious Experience: A Study in Human Nature (Being the Gifford Lectures on Natural Religion Delivered at Edinburgh in 1901-1902),* The Modern Library, New York, NY.

Johnson, David Martel. 1988. "'Brutes Believe Not,'" *Philosophical Psychology,* Vol. 1, No.3, pp.279-94.

——. 1990. "Can Abstractions be Causes?" *Biology and Philosophy,* Vol. 5, pp.63-77.

——. 1997. "Taking the Past Seriously: How History shows that Eliminativists' Account of Folk Psychology is Partly Right and Partly Wrong," pp.366-75 in Johnson and Erneling, 1997.

——. 2003. *How History Made the Mind: The Cultural Origins of Objective Thinking,* Open Court. Chicago.

——. 2005. "Mind, Brain, and the Upper Paleolithic," pp.499-510 in Erneling and Johnson, 2005.

Johnson, David Martel and Christina E. Erneling (editors). 1997. *The Future of the Cognitive Revolution,* Oxford University Press, New York, NY.

Keller, Helen. 1903/1996. *The Story of My Life,* Dover Publications, Mineola, NY.

——. 1904/1910. *The World I Live In,* The Century Co., New York, NY.

King, Barbara J. 1994. *The Information Continuum: Social Information Transfer in Monkeys, Apes, and Hominids,* School of American Research Press, Santa Fe, NM.

——. 2001. *Roots of Human Behavior,* THE TEACHING COMPANY, Chantilly, VA.

——. 2002. *Biological Anthropology: An Evolutionary Perspective, Parts I and II,* THE TEACHING COMPANY, Chantilly, VA.

Klein, Richard G. with Blake Edgar. 2002. *The Dawn of Human Culture,* John Wiley & Sons, New York, NY.

Jablonski, N.G. and L.C. Aiello (Eds.). 1998. *The Origin and Diversification of Language,* Memoirs of the California Academy of Sciences, no.24, San Francisco, CA.

Lange, Karen E. 2002. "Wolf to Woof: The Evolution of Dogs," *National Geographic,* January, pp.2-31.

Leaky, Richard. 1994. *The Origin of Humankind,* Basic Books, New York, NY.

Leonard, Jonathan Norton and the editors of Time-Life Books. 1973. *The First Farmers,* Time-Life Books, New York, NY.

Lewis-Williams, David and David Pearce. 2003. *Inside the Neolithic Mind: Consciousness, Cosmos and the Realm of the Gods,* Thames and Hudson, New York, NY.

Lieberman, Philip. 1991. *Uniquely Human: The Evolution of Speech, Thought, and Selfless Behavior,* Harvard University Press, Cambridge, MA.

——. 2000. *Human Language and Our Reptilian Brain: The Subcortical Bases of Speech, Syntax, and Thought,* Harvard University Press, Cambridge, MA.

Lloyd, G.E.R., 1970. *Early Greek Science: Thales to Aristotle,* Chatto and Windus, London, U.K.

Mayr, Ernst. 1991. *One Long Argument: Charles Darwin and the Genesis of Modern Evolutionary Thought,* Harvard University Press, Cambridge MA.

Mellars, Paul. 1998. "Neanderthals, Modern Humans and the Archaeological Evidence for Language," pp.89-115 in Jablonski and Aiello, 1998.
Merton, Thomas. 1996a. *A Search for Solitude. Journals*, V.3. edited by Lawrence S. Cunningham, HarperSanFrancisco, San Francisco, CA.
——. 1966b. *Conjectures of a Guilty Bystander*, Doubleday, New York, NY.
McAuliffe, K. 2009. "Are We Still Evolving?" *Discover*, March, pp.51-8.
Midgley, Mary. 1998. "One World, but a Big One," Chapter 14 in Rose (1998).
Mithen, Steven. 1996. *The Prehistory of the Mind: The Cognitive Origins of Art, Religion and Science*, Thames and Hudson, London, U.K.
Moore, A.W., 1991. *The Infinite*, Routledge, London, U.K.
Morris, Desmond. 1967. *The Naked Ape: A Zoologist's Study of the Human Animal*, McGraw-Hill, New York, NY.
Ockham, William. 1962. *Ockham: Philosophical Writings*, edited and translated from the Latin by Philotheus Boehner, Nelson, New York, NY.
Pepperberg, Irene. 2008. *Alex and Me*, HarperCollins Publishers, New York, NY.
Peterson, Roger Tory. 1964. *A Field Guide to the Birds*, Houghton Mifflin Company, Boston, MA.
Phillips, Angus. 2002. "A Love Story: Our Bond with Dogs," *National Geographic*, January, pp.12-31.
Pica, P.; C. Lemer; V. Izard; S. Dehaene. 2004. "Exact and Approximate Arithmetic in an Amazonian Indigene Group," *Science*, Vol.306 (15 October), pp.499-503.
Pinker, Steven. 1994. *The Language Instinct: How the Mind Creates Language*, Harper-Collins, New York, NY.
——. 1997. *How the Mind Works*, W.W. Norton & Co., New York, NY.
Jane Poynter. 2006. *The Human Experiment: Two Years and Twenty Minutes inside Biosphere 2*, Thunder's Mouth Press, New York, NY.
Quammen, David. 2004. "Was Darwin Wrong?" *National Geographic*, November, pp.2-35.
Renfrew, Colin. 2008. *Prehistory: The Making of the Human Mind*, The Modern Library, New York, NY.
Robinson, Janie. 2009. "The Spirits of the Vikings: Following the Trail of Norse Explorers on Remote Newfoundland Peninsula," *Toronto Star*, May 28, pp.T1-2.
Rose, Steven (Ed.). 1998. *From Brains to Consciousness?* Princeton University Press, Princeton, NJ.
Ruse, Michael. 1999. *Mystery of Mysteries: Is Evolution a Social Construction?*, Harvard University Press, Cambridge, MA.
Ryle, Gilbert. 1949. *The Concept of Mind*, Hutchinson & Co., London, U.K.
Sartre, Jean Paul. 1948/1964. Extract from *Existentialism and Humanism* in Morton White, 1964.
Savage, Candace. 1995. *Bird Brains: The Intelligence of Crows, Ravens, Magpies and Jays*, Greystone Books, Vancouver and Toronto.
——. 2005. *Crows: Encounters with the Wise Guys*, Greystone Books, Vancouver, Toronto, and Berkeley.
Savage-Rumbaugh, Sue; Stuart G. Shanker; Talbot J. Taylor. 1998. *Apes, Language, and the Human Mind*, Oxford University Press, New York, NY.
Searle, John R. 1992. *The Rediscovery of the Mind*, MIT Press, Cambridge, MA.
——. 2002. "End of the Revolution," *The New York Review of Books*, Vol.49, No.3, February 28, 2002.

Shreeve, Jamie. 2010. "The Evolutionary Road," *National Geographic*, July, pp.34-67.
Stamos, David N. 2003. *The Species Problem: Biological Species, Ontology, and the Metaphysics of Biology*, Lexington Books, New York, NY.
Strauss, Stephen. 2004. "Life without Numbers in a unique Amazon Tribe," *The Globe and Mail*, Friday, August 20, p.A3.
Stringer, Christopher and Robin McKie. 1996. *African Exodus: The Origins of Modern Humanity*, Henry Holt, New York, NY.
Tattersall, Ian. 1998. *Becoming Human: Evolution and Human Uniqueness*, Harcourt Brace & Co., San Diego, CA.
——. 1999. *The Last Neanderthal: The Rise, Success, and Mysterious Extinction of Our Closest Human Relatives* (Revised edition), Westview Press, Boulder, CO.
——. 2000. "Once We were Not Alone," *Scientific American*, January, pp.56-62.
Thompson, W. 1996. *Coming into Being: Artifacts and Texts in the Evolution of Consciousness*, St. Martin's Press, New York, NY.
Thorne, Alan G. and Milford H. Wolpoff. 1992. "The Multiregional Evolution of Humans," *Scientific American*, April, pp.76-83.
United States Navy website. (Navy. Mil The Official Website of the United States Navy. www.navy.mil/)
Wade, Nicholas. 2010. "From the big, bad wolf comes Fido and Fluffy: Researchers analyzed thousands of genetic markers to find that today's pet dogs descend from Middle Eastern grey wolves," *The Globe and Mail*, Thursday, March 18, p.A3.
White, Morton. 1964. *The Age of Analysis: Twentieth Century Philosophers*, Mentor Books, Toronto.
Whiten, Andrew and Christopher Boesch. 2001. "The Culture of Chimpanzees," *Scientific American*, January, pp.61-7.
Wilson, Edward O. 1975/1980. *Sociobiology: The Abridged Edition*, Harvard University Press, Cambridge, MA.
——. 1978/2004. *On Human Nature*, Harvard University Press, Cambridge, MA and London, U.K.
——. 1992. *The Diversity of Life*, Harvard University Press, Cambridge, MA.
——. 1998. *Consilience: The Unity of Knowledge*, Alfred A. Knopf, New York, NY.
Wiseman, Boris. 1997. *Lévi-Strauss for Beginners*, Icon Books, Cambridge, U.K.
Wong, Kate. 2005. "The Littlest Human," *Scientific American*, February, pp.56-65.
Wong, Kate and Viktor Deak. 2009. "The Human Pedigree," *Scientific American*, January, pp.60-3.
Zimmer, Carl. 2005. *Smithsonian Intimate Guide to Human Origins*, Collins, New York.

Name Index

Y

Z

Subject Index

G

H

I

K

L

M

U

W